C MATERIALS SCIENCE OF ONCRETE VII

T0328467

MATERIALS SCIENCE OF CONCRETE VII

EDITED BY FRANCIS YOUNG
JAN SKALNY

Published by
The American Ceramic Society
735 Ceramic Place
Westerville, Ohio 43081
www.ceramics.org

The American Ceramic Society
PO Box 6136
Westerville, Ohio 43086-6136

ISBN: 1-57498-210-9

Library of Congress Cataloging-in-Publication Data
A CIP record for this book is available from the Library of Congress.

For more information on ordering titles published by The American Ceramic Society or to request a publications catalog, please call (614) 794-5890 or visit our online bookstore at <www.ceramics.org>.

Contents

Prediction of a Portland Cement's Properties from Its Chemical and Mineralogical Composition

Danielle Sorrentino, François Sorrentino, and Ellis Gartner

It was not until almost 100 years after the invention of portland cement that Bogue published his well-known formulae that allowed the producer to estimate the potential phase composition of the product from its chemical composition and thus to calculate the desired kiln feed composition. Producers, users, and researchers are all interested in this subject, but each of them for somewhat different reasons.

This chapter deals with the origins of these formulae, their uses and limitations, and improvements that have been made in the seven decades since their promulgation. H. F. W. Taylor's semi-empirical approach has provided the most practically useful improvements.

The formulae themselves were never intended to give a precise prediction of clinker composition, since it was well understood by Bogue that actual clinker phase composition depends on both the thermal history and the presence of unreactive materials and minor components in the raw feed. This paper highlights the importance of both of these phenomena and explains why the actual performance of portland cement cannot easily be correlated with its "Bogue-calculated" composition. We deal in detail with only what we consider to be the most important minor elements for modern-day clinker production, that is, the alkali metals, sulfur, magnesium, and phosphorus.

Introduction: The Need to Predict Cement Properties

Ever since the development of portland cement technology, both producers and users have felt the need to be able to predict the practical ("usage") properties of the cement by some type of analytical procedure that can be conducted on the product, preferably before its use, for example, in concrete. For the producer, this is felt to be essential as a low-cost means of controlling the quality of the product leaving the factory. For the user, it should serve as a low-cost means of not only assessing the probable quality of the product prior to purchase, but also comparing multiple offers of similar products from different suppliers with regard to their suitability for the job in question.

For the producer, the most important requirement is to be able to rapidly correct an anomaly in clinker production. Because of the high production rate of a modern portland cement plant, the producer risks having to waste several hundred tons of clinker if a serious defect (for example, excessive free lime content) goes uncorrected for just a few hours.

Table I. Variation of C₃S content in a clinker at constant percentage CaO

% C_3S	% CaO	LSF	A/F	S/(A+F)
66	70	94.83	2.44	5.40
68	70	95.48	1.94	5.03
70	70	96.14	1.60	4.70
72	70	96.81	1.36	4.41
74	70	97.48	1.18	4.14
76	70	99.51	1.07	3.36

Of course, for the user, the resulting cement must also satisfy the norms, which generally require a certain range of normalized mortar compressive strengths at ages up to 28 days. Without a reliable method for predicting these strengths from rapid clinker and cement analyses, the cement producer would risk losing at least 28 days' worth of production in the case of a serious anomaly that could only be detected by the standard mortar test.

Clearly, then, the producer must have rapid performance prediction methods at his disposal. The simplest of these would be a measure of total clinker CaO content. But is this sufficient? Presumably not, since one can vary cement properties while holding clinker CaO content constant. If one varies the silica ratio [S/(A+F)] and the alumina ratio (A/F) simultaneously, one can easily show that the predicted C_3S content varies at constant CaO (see Table I).

When moving from lower to higher C_3S content, the following general improvements are observed:
- Burnability is better, owing to a lower A/F.
- Sulfate resistance increases with decreasing A/F (decreasing C_3A).
- The grinding energy is reduced when C_3S increases at the expense of C_2S.
- Strengths increase with increasing C_3S.

Given that the analytical methods available to both cement producers and users in the nineteenth century were very limited by modern standards, it is hardly surprising that it took more than a half-century before it was realized that the most important mineral phases in portland cements are alite and belite, and that early age strength development in concrete depends mainly on the content of alite in the cement. The actual compositions of the main clinker phases were not known with much certainty until long after they were first identified by optical microscopic examination.

Some of the fundamental work necessary to establish the high-temperature phase equilibria of this system was done in R. H. Bogue's research group at the National Bureau of Standards in Washington, D.C., under the Portland Cement Association Fellowship in the late 1920s and early 1930s. In 1929, Bogue[1] published a simple method of calculating the phase composition of an idealized clinker based on its chemical analysis and the assumption that it was composed of the four principal phases: C_3S, C_2S, C_3A, and C_4AF. This led to a breakthrough in the way that both producers and users understood the product. Soon after this, Lea and Parker[2] published crucial phase stability data for the subsystem C-C_2S-$C_{12}A_7$-C_4AF, which showed that things were a bit more complicated than had been assumed in the earlier analyses but did not change the basic concepts.[3]

Surprisingly, these breakthroughs in fundamental understanding did not occur until almost a century after the first commercial use of a product that can reasonably be called portland cement. Apparently, although the need had presumably been there all along, everyone managed to get along quite nicely without a phase prediction method before that, because they simply tested the products on offer under practical conditions simulating their actual needs. In the practical world, it is essentially the issue of cost that differentiates between the analytical approach and the empirical testing approach. However, in the modern world, performance testing is usually considerably more expensive than instrumental analytical procedures and, of course, takes far longer, which results in a proportionately higher financial risk in the case where a bad product is produced. On the other hand, the issue is not at all the same for a true scientist (typified by Hal Taylor) whose primary objective is to understand how cement works, or for the inventor who wishes to go one step further and use this understanding develop something better, or for the entrepreneur who wishes to go one step further still and commercialize the resulting invention.

Process: How to Calculate the Desired Kiln Feed Composition

The Bogue formulae represent a simple linear transformation of the elemental analysis of the clinker, or, in the case of raw mix control, of the raw mix minus the ignition loss. Clearly, this takes no account of the actual state of combination of the elements in the clinker. The very minimum process variable required to be taken into account is the measurement of actual clinker free lime. One can also go one step further in measuring the amount of sulfate and its state of combination (e.g., alkali sulfate or calcium sulfate). But the development of new methods of calculation has not removed the confusion that exists regarding their objectives.

The first objective is to calculate the raw mix composition required to give an appropriate clinker composition compatible with efficient quarry utilization. This requires a predictive approach, typically based on moduli such as the silica ratio, the alumina ratio, and the lime saturation factor.

A second objective is to control the clinkering process. For this, we do not need to measure the full clinker composition. The simplest method is the measure of clinker bulk density (unit weight), but the most widely accepted analytical method is free lime. The principal requirement of such methods is rapidity, in order to avoid lost production.

However, when a problem arises in product quality or in processing, it is necessary to have a more accurate diagnosis. In this case, a thorough clinker analysis is required, and optical microscopy is usually found to be one of the most effective approaches.

When planning the production of a totally new cement, or simply the optimization of an existing cement for specific performance requirements, one would ideally like to know the fundamental relationships between clinker composition and cement performance. In order to develop such relationships to an advanced level, accurate, precise, and repeatable analytical procedures are required for the clinker phases.

Experience shows that there are genuine relationships between clinker phase composition and certain key performance parameters. For example, in the case of standardized mortar tests (at room temperature and at ages up to 28 days) compressive strength correlates moderately well with alite content, all other independent variables being held roughly constant. However, other independent variables, such as fineness, can have just as great an influence. Thus, while it is clear that a precise clinker phase analysis is necessary for reliable prediction of key cement performance properties, it is evidently by no means sufficient, and it is not always clear which cement phases play the most important role in any given performance aspect. Thus, the question remains: Which clinker phases need to be measured, and with what precision, in order to obtain the maximum useful information for cement performance prediction?

Numerous techniques have been applied to clinker phase analysis, including optical microscopy, selective dissolution, SEM (with or without X-ray spectroscopy), XRD (with or without Rietveld analysis), and so on.

While great progress has been made in the accuracy and repeatability of the analytical procedures, the results have in general been rather disappointing in terms of cement performance prediction. There appear to be two main reasons for this:

1. The principal property of interest has almost always been standard mortar strength, which is not an intrinsic property of a cement.
2. The results of many of the analytical methods are parameters that may have little to do with many of the performance characteristics of interest.

The Chemistry of Cement Manufacture

As is well known, portland cement clinkers are generally manufactured by heating a finely ground mixture of limestone, clay (or other aluminosilicate minerals), and iron oxide to temperatures usually in excess of 1400°C, and cooling the resulting clinker rapidly in air. The chemical composition is always given in oxide notation, since, under normal manufacturing conditions, all of the most important elements in clinker occur with oxygen in the stoichiometry of their highest oxidation states. This is a very important simplification, but it has its limits. The principal element most likely to diverge from this approximation is iron, for which a significant fraction of the ferrous oxidation state (Fe^{II}) can be produced at clinkering temperatures, even within the range of oxygen partial pressures found in normal portland cement kilns (which typically are of the order of 1% excess O_2 in the kiln gases, but with some local variations around the flame). However, Fe^{II} tends to be reoxidized to Fe^{III} on cooling, so most clinkers contain very little Fe^{II} (apart from white cements, where reducing conditions are often deliberately used to improve the color).

Given that the five major elements in portland clinkers are always (in order of abundance) oxygen, calcium, silicon, aluminium, and iron, the minimum number of oxide components required to specify a clinker is four: CaO, SiO_2, Al_2O_3, and Fe_2O_3. If it is assumed that equilibrium is reached at the maximum temperature attained during clinker formation, the phase rule dictates that the maximum number of condensed phases that can normally be present is also four. The phase diagrams developed by Bogue's group,[3] Lea and Parker,[2] and others show that at temperatures above about 1250°C, the four Bogue phases — C_3S, C_2S, C_3A, and C_4AF — form a stable assemblage. It is thus relatively easy to conclude that the phase composition of a pure ideal portland clinker can be predicted from the proportions of these four components by a simple set of four linear simultaneous equations. That is the concept of the classic Bogue formulae,[1] as shown here:

$$C_3S = 4.0710\,CaO - 7.6024\,SiO_2 - 6.7187\,Al_2O_3 - 1.4297\,Fe_2O_3 \quad (1)$$

$$C_2S = -3.0710\,CaO + 8.6024\,SiO_2 + 5.0683\,Al_2O_3 + 1.0785\,Fe_2O_3 \quad (2)$$

$$C_3A = 2.6504\,Al_2O_3 - 1.6920\,Fe_2O_3 \qquad\qquad\qquad (3)$$

$$C_4AF = 3.0432\,Fe_2O_3 \qquad\qquad\qquad\qquad\qquad (4)$$

It is thus possible to calculate the phase composition of a clinker from its chemical composition. However, in practice, two difficulties are encountered :

1. The observed phase composition is never exactly that predicted by the Bogue formulae.
2. The calculated phase composition doesn't correlate well with cement performance.

These two difficulties are not necessarily related to each other. The idea that an increase in the accuracy of phase measurements will improve the correlation with practical performance assumes that a correlation must exist, but this is not fully proven for all properties of cement. In other words, even if we increase the accuracy of the estimation of alite content, we are not certain to improve the accuracy of strength prediction. Attempts to resolve these issues have been the object of more than half a century of research.

How Are Raw Mix Compositions Chosen?

The raw mix composition is chosen by the cement producer to allow efficient cement manufacturing based on his raw materials and fuel sources and the type of kiln system. How does he make the choice? Generally, two parameters are used for controlling the clinkering process:

1. The alumina ratio (AR) — (A/F); usually kept between 1.3 and 2.
2. The silica ratio (SR) — [S/(A+F)]; usually kept between 2.3 and 3.1.

These moduli are chosen mainly to ensure that good strong clinker particles form and persist without too much softening in the burning zone. In the absence of such behavior, the kiln would not run well, owing to either excessive coating formation or excessive dust cycles. Of course, a different process, for example, a melt process, would necessitate different parameters or different values. For example, in the steelmaking industry the impurities associated with iron ore (mainly silica and alumina) must be neutralized by lime, but the process requires a low-viscosity liquid, which leads the steel industry to produce a slag and not a clinker.

In order to make a good portland clinker, it is necessary to include sufficient lime to combine with all the silica, alumina, and iron oxide and produce a high percentage of alite. Hence, a "lime saturation factor" is required. There are several different LSF indices in common use: the Lea and Parker LSF, the Kühl KSG, the Lafarge delta, and so on, as explained below.

The maximum acceptable content of CaO for burning a portland cement clinker can be derived from Lea and Parker phase diagram[2] at the boundary between two regions: the primary C_3S phase volume, where alite coexists with liquid phase, and the free-lime region where free lime (calcia, CaO) coexists with liquid phase. Lea and Parker derived their lime saturation factor by making use of the invariant (peritectic) point P3 $(C_3S/C_3A/CaO/melt)$ at 1470°C in the iron-free system, which has a liquid phase composition containing 59.7% CaO, 32.8% Al_2O_3, and 7.5% SiO_2.[4] They found that the isotherm separating the primary phase boundaries for C_3S and for CaO in the four-component system could be approximated in the most important part of the composition range by a plane bounded by P3, C_3S, and C_4AF, and which was therefore defined by the following equation (in terms of mass proportions):

$$CaO = 2.80\,SiO_2 + 1.18\,Al_2O_3 + 0.65\,Fe_2O_3 \qquad (5)$$

Thus, values of CaO higher than those given by the formula represent points in which some CaO will always be present at equilibrium. The ratio of the actual CaO content to the expression on the right-hand side of Eq. 5 was called the lime saturation factor (LSF), because, in theory, a raw mix having an LSF greater than 100% will yield free lime in the final clinker irrespective of the degree of mixing, the fineness of the raw mix, and the degree of burning:

$$LSF = 100\,CaO\,/\,(2.80\,SiO_2 + 1.18\,Al_2O_3 + 0.65\,Fe_2O_3)$$

Note that Eq. 5 can be transformed into molar proportions (C, A, F, S,) as shown below:

$$C = 3\,S + 2.147\,A + 1.853\,F \qquad (6)$$

This indicates that the plane intersects the C-S axis at the point of composition C_3S, the C-A axis at the point $C_{2.147}A$, and the C-F axis at the point $C_{1.853}F$.

However, the equation is not necessarily valid at these extremes — certainly not at the alumina- and iron-rich extremes, in any case, since neither of the two compositions ($C_{2.147}A$ or $C_{1.853}F$) represents a stable phase. On the other hand, it is valid for compositions of intermediate alumina ratio, and it can be easily seen that, for A = F, it gives the correct lime saturation factor for C_4AF. Lea and Parker considered that the equation was only valid for A/F molar ratios of one or more (above 0.64 on a mass basis), but Taylor[4] remarks that later data by Swayze showed that it could be extended much further into the iron-rich composition range with reasonable precision.

One should be careful not to confuse Lea and Parker's LSF with the Bogue equations, since it is designed to take into account the actual melt phase composition at typical clinkering temperatures, and this melt is actually lime-deficient with respect to C_3A in alumina-rich compositions. If one assumes instead that the simple molar proportions for the four solid clinker phases, as used in the Bogue equations, are valid at clinkering temperatures, one can obtain a different equation for lime saturation, simply by setting the percentage of C_2S at 0 in Eq. 2. In mass terms:

$$CaO = 2.80\, SiO_2 + 1.65\, Al_2O_3 + 0.35\, Fe_2O_3 \tag{7}$$

Or, in molar terms:

$$C = 3\,S + 3\,A + F \tag{8}$$

The factor defined by Eq. 7 is often referred to as Kühl's (lime) saturation grade (KSG):

$$KSG = 100\, CaO\, /\, (2.80\, SiO_2 + 1.65\, Al_2O_3 + 0.35\, Fe_2O_3)$$

As is evident from the molar formulae (Eqs. 6 and 8), both the LSF and the KSG have the same value at A = F, but diverge for molar A/F ratios different from unity. The KSG intersects the C-A axis at the point C_3A, which suggests that it is probably better than the LSF for very low-iron clinkers; but it intersects the C-F axis at the point CF, which represents a compound that is not phase compatible with C_3S and either C or C_2S. The ferrite phase found in very iron-rich clinkers is close to the C_2F end of the C_4AF-C_2F solid solution, which is much better approximated by the Lea and Parker LSF formula. It therefore appears that the LSF is actually better suited to

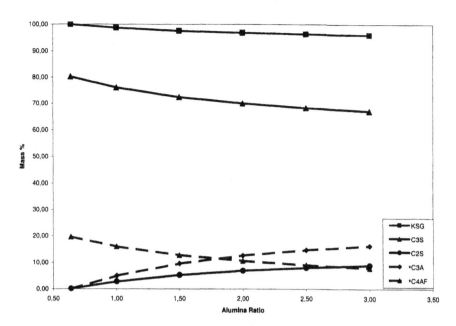

Figure 1. KSG and Bogue composition calculated for clinkers with SR = 2 and LSF = 100.

clinkers with alumina ratios below about 0.64, whereas the KSG may be slightly better suited to clinkers with alumina ratios above that value. However, this is not necessarily true, for the reasons given above. In practice, the LSF should be more accurate over most of the range actually encountered in real clinkers, whereas setting the KSG to 100% will tend to lead to mixtures with slightly too much lime to be completely combined at normal clinkering temperatures (for alumina ratios above 0.64). This is mainly because the solid and liquid phases usually do not have time to fully reequilibrate on cooling to complete solidification, under which conditions the four Bogue phases should coexist.

The discrepancy between the LSF and the KSG can be shown by calculating the Bogue compositions for different hypothetical "pure" clinkers (C + A + F + S = 100%). This has been done in Fig. 1 for clinkers with a fixed silica ratio of 2 and LSF of 100%, over a range of alumina ratios. It can be seen that there is no simple relationship between the various factors.

It should be noted that the calculations here were only made for alumina ratios of 0.64 and higher. This is because the simple Bogue formulae are not appropriate for lower AR values, as they predict a negative C_3A content. We can, however, make use of Lea and Parker's data to go to lower

AR values, by assuming that the ferrite phase composition will lie between C_4AF and $C_{1.853}F$. It is thus possible to rewrite the Bogue formulae as follows for AR < 0.64:

$$C_3S = 4.06\,CaO - 7.57\,SiO_2 - 4.78\,Al_2O_3 - 2.65\,Fe_2O_3 \qquad (9)$$

$$C_2S = -3.06\,CaO + 8.57\,SiO_2 + 3.60\,Al_2O_3 + 1.99\,Fe_2O_3 \qquad (10)$$

$$Ferrite = 2.18\,Al_2O_3 + 1.65\,Fe_2O_3 \qquad (11)$$

Alternatively, it can be assumed that the ferrite phase lies between C_4AF and C_2F, as observed for the solid solution in the C-A-F system; but this will increase the risk of excess free lime.

Analytical Methods

It is not our objective here to give a detailed description of all of the analytical methods that can be used for cements and clinkers, but simply to remind the reader that it is important to be aware of their limitations. A review of the techniques currently used for quantitative clinker phase analysis has been published in a recent volume of this series.[5]

X-Ray Diffraction

Quantitative powder X-ray diffraction techniques have been widely used on cements since the pioneering work of Brunauer and Copeland.[6] The early studies generally made use of peak height measurements wherever characteristic diffraction peaks could be found without excessive overlap. This approach was soon improved by using peak area measurements and by comparing the diffraction patterns of real cements with those of pure phases (and their mixtures) synthesized in the laboratory. The principal limitation of the technique has always been peak overlap, which is exacerbated by the fact that many of the phases present in portland cements have low crystal symmetries (monoclinic or triclinic) and thus exhibit a large number of weak diffraction peaks. The first approach to reducing the problem of peak overlap was to develop selective dissolution techniques for specific clinker phases.[7] More recently, sophisticated algorithms have been developed that allow for the calculation of the complete powder diffraction pattern of any phase, given its crystal parameters.[8] This allows direct computation of the powder diffraction patterns of complex mixtures of known phases, and also allows the inversion of the problem in the case where the crystalline structures of all of the phases in the mixture are known.[5]

In addition to solving the problem of peak overlap, these newly developed algorithms allow for a much better correction in the case of unavoidable preferred orientations of individual phases, a problem that had previously been very difficult to treat because no matter how carefully one tries to avoid it, particles with a nonspherical shape always tend to orient in preferred directions relative to the plane of the sample surface. There remain, however, two intrinsic limitations: the sensitivity (signal/noise ratio and angular resolution) of the diffractometer, and the requirement that all of the phases in the assemblage must have a high degree of crystallinity. Truly amorphous phases can be treated only by very empirical methods, for example, by the subtraction of a typical pattern obtained for the "pure" amorphous phase. However, crystalline phases with significant peak broadening due to specific disordering or small crystallite size can now be treated to some extent by sophisticated theoretical approaches. And the sensitivity of diffractometric equipment has been greatly enhanced by the development of better detectors and more powerful monochromatic X-ray sources.

Selective Dissolution

Several techniques have been developed for preferentially dissolving specific phases in portland cements without greatly changing some of the other phases, thus leaving a residue enriched in the phases of interest. All such techniques have been developed and validated empirically; they can never be perfect, but they do nevertheless greatly facilitate the study of the less abundant phases that are commonly encountered in the mixtures.

The first use of selective dissolution was made in the development of chemical techniques for the quantitative estimation of the free lime content of clinkers.[9]

Later, it was found that solutions of salicylic or maleic acid in anhydrous ethanol can dissolve the principal calcium silicate phases of clinker (alite and belite) and free lime, without much attack on the aluminate, ferrite, or sulfate phases. If desired, the residue could be further treated with a concentrated sucrose solution to remove the aluminate and sulfate phases, leaving only the ferrite phase.

The use of a solution of sucrose in concentrated potassium hydroxide solution allows one to directly dissolve the interstitial phases of clinker (aluminate and ferrite) with little effect on the silicate phases.

Selective dissolution techniques have also been developed to estimate the state of division calcium sulfates in a cement by a difference method.[10] Traditionally, the phases or mixtures remaining after selective dissolution are simply weighed, analyzed chemically, or studied by quantitative XRD.[11,12]

Optical Microscopy

The quantitative phase composition of clinker can also be determined by optical microscopy. This is generally done by means of a polished section viewed in reflected light. A wide variety of phase-specific etching techniques have been developed to highlight specific phases and thus render them identifiable.[13] The identifiable phases are quantitatively determined by point counting, leading directly to the relative volume proportions of the counted phases. Corrections must then be made for phase densities and the presence of voids or other uncharacterized phases in order for these numbers to be converted into weight fractions.

Clearly, the accuracy of these optical techniques depends on the skill of the microscopist in preparing representative etched polished sections and identifying the individual phases. This work is also rendered very difficult if the phases are present in very small regions that cannot be resolved optically, as is often the case with the interstitial (aluminate and ferrite) phases. The precision is determined essentially by simple counting statistics, being $\sqrt{n}$ where n is the number of counts per phase. It can be seen, therefore, that it is necessary to count about 10 000 points in order to have a precision of about 1%. For all of the above reasons, together with the fact that it has proven very difficult to reliably automate the process and the high cost of skilled labor, the optical microscopic approach has fallen out of favor in recent times.

The optical point counting technique has been applied to a large number of industrial clinkers, and a detailed statistical analysis of the results combined with a comparison with clinker elemental composition have led to the development of clinker phase composition formulae that can be used in a predictive manner as an alternative to the Bogue formulae[14]:

$$\text{alite} = 5.17\,CaO^* - 7.62\,SiO_2 - 5.79\,Al_2O_3 - 3.12\,Fe_2O_3 + 4.27\,MgO + 12.8\,R_1 - 71.42$$

$$\text{belite} = -4.59\,CaO^* + 9.21\,SiO_2 + 4.87\,Al_2O_3 + 2.74\,Fe_2O_3 - 4.70\,MgO - 16.73\,R_1 + 89.42$$

$$\text{aluminate} = 2.02\,Al_2O_3 - 1.62\,Fe_2O_3 - 1.52\,MgO + 1.49\,R_1 + 1.49$$

$$\text{ferrite} = 2.69\,Fe_2O_3 + 1.61\,MgO - 4.62$$

$$\text{periclase} = 0, \text{ for } MgO < 2\%$$
$$= MgO - 2\%, \text{ for } MgO > 2\%$$

The value of CaO used in above calculations is a corrected value, as follows:

$$CaO^* = \text{total CaO} - \text{free CaO} - 1.27\,CO_2 - 0.7\,SO_3$$
$$- 0.85 \text{ soluble } K_2O - 1.29 \text{ soluble } Na_2O$$

R_1 (lattice alkalis) is total alkalis dissolved in the alite, belite, aluminate, and ferrite phases, expressed as oxides and as a percentage of total clinker mass. (Lattice alkalis represent all of the alkalis that are not immediately water soluble, that is, not present as sulfates.)

These formulae can also be written as corrections to the Bogue-calculated phase percentages:

$$\text{alite} = C_3S_{Bogue} + \{9.71 - 2.40\,Fe_2O_3 + 3.65\,MgO + 8.45\,R_1\}$$

$$\text{belite} = C_2S_{Bogue} + \{2.40\,Fe_2O_3 - 3.55\,MgO - 10.91\,R_1 - 2.9\}$$

$$\text{aluminate} = C_3A_{Bogue} + \{4.75\,R_1 - 1.14 - 2.03\,MgO\}$$

$$\text{ferrite} = C_4AF_{Bogue} + \{1.75\,MgO - 5.84\}$$

It is interesting to observe that the required corrections to the Bogue formulae lie in the direction predicted by a solid-solution model in which significant amounts of alkalis, Mg, and Al are taken into the alite (and substitute mainly for calcium). Thus, the actual alite content in a well-burned clinker is almost always significantly higher than that predicted by the Bogue formulae, whereas the amounts of belite and aluminate tend to be lower. The only factor that tends to increase the content of aluminate phase is the presence of lattice alkalis.

Optical microscopy can also be used in the transmitted light mode for assessing the mean crystallite size and optical properties of the alites and belites. Clinker thin sections can be used, but are very tedious to prepare. Powder mounts are very easy to prepare (the cement itself can be used) and permit the analysis of those small grains that are essentially monocrystalline. This approach was developed principally by Ono[15] and proved useful in the relative evaluation of clinkers from a single kiln, that is, in cases where the raw materials and clinker compositions did not vary widely. In addition to allowing control of kiln burning conditions, this approach was also used successfully in some cases to predict cement strength development in standard mortars, and thus to correlate burning conditions with cement performance. However, it was found to be difficult to apply these concepts widely, and in many cement plants they did not prove particularly helpful.

Scanning Electron Microscopy

It is also feasible to use a scanning electron microscope (SEM) in backscattered electron mode to quantitatively analyze the phase proportions by means of polished sections, but one encounters difficulties similar to those encountered with optical microscopy, although the image analysis is somewhat easier to computerize since all SEMs already have electronic data acquisition systems. No etching is used, and the phases can be distinguished either by mean atomic number contrast or by elemental composition as determined by X-ray fluorescence. In the first case, the spatial resolution can be somewhat higher than with optical microscopy, but it is not easy to distinguish many of the phases, except for the ferrite phase, because it is the only one with a very significantly different mean atomic number. In the second case, the spatial resolution is no better than with optical microscopy and counting times can be quite long for the X-ray analysis.

However, the possibility of using X-ray fluorescence to accurately determine the elemental composition of crystalline phases also allows SEM (or, even better, an electron microprobe equipped with a wavelength-dispersive X-ray spectrometer) to be used to estimate the phase proportions by a completely different manner. This is the approach first used by Yamaguchi and Takagi.[11] Unfortunately, the resolution is limited to crystallites larger than a few microns in apparent area (as much as with optical microscopy), since in this case the elemental analysis is otherwise too strongly influenced by the adjacent phases.

All of the above methods can be used separately or, even better, in combination to obtain even more information than the simple phase proportions. Table II summarizes the different approaches and the types of information that they can be expected to give.

Differences between Calculated and Observed Clinker Phase Compositions

As was pointed out clearly by Bogue,[1] small errors in elemental analysis are magnified by his formulae. It was for this reason that he counseled that the calculated phase compositions should not be given to more than one decimal place, a recommendation that is, sadly, frequently ignored by many analytical laboratories. Typical error ranges of the oxide analyses of cements or clinkers by current instrumental techniques are as follows: CaO ±0.5%; SiO_2 ±0.5%; Al_2O_3 ±0.2%; Fe_2O_3 ±0.2%. These errors can give rise to very significant variations in calculated compound composition, especially if the effects accidentally reinforce one another, as is shown in Table III, an example in which the lime and silica errors reinforce each other to the maximum extent possible around the mean values given in the middle row. It can be seen that, in reality, it is probably not wise to give any decimal place to the calculated percentage phase compositions; integer values should suffice given the actual precision of most analyses. In any case, we surely do not know how to interpret performance differences with any greater precision than that.

The basic Bogue equations are, in principle, only truly applicable in the ideal case where $C + S + A + F = 100\%$ of the mass of the clinker. However, this is never the case in practice. The remainder of the clinker mass consists of other components not included in the calculation, and it is the question of how to treat these other components that gives rise to the greatest difficulty. We don't know the full phase diagram in the presence of even the most common minor components (MgO, SO_3, K_2O, Na_2O, FeO). Clearly, even if we did know the full phase diagram in the presence of these additional components, it would be difficult to manipulate (it would have to be computer-modeled) and it would still allow us to only estimate the equilibrium compositions at any given temperature. Note, also, that there is a tendency for the contents of additional elements to be quite significant in some modern clinkers, especially when industrial by-products are used as fuel or alternative raw materials. Elements that are quite commonly found in significant quantities include Zn, Ti, P, and Mn. The current method of dealing with these complexities is to just make modifications to the Bogue formulae for the principal phases, but this is oversimplistic.

It should be noted that Bogue originally considered that essentially all of

Table II. The different approaches and the types of information they provide

Method	Input data required (optional data)	Computational method used	Principal application	Supplemental data outputs
Bogue calculation (kiln feed)	Bulk chemical analysis of kiln feed	Matrix algebra	Prediction of clinker phase composition	None
Bogue calculation (cement or clinker)	Bulk chemical analysis (*free lime, insoluble residue, and soluble alkalis*)	Matrix algebra	Cement or clinker phase calculation	Can give estimates of other phases in addition to the four principal phases
QXRD (simple)	Phase standards for comparison (*selective dissolution residue proportions*)	Peak height or area comparisons	Cement or clinker phase analysis	Polymorphism of principal phases
QXRD (Rietveld)	Rietveld data files (*selective dissolution residue proportions*)	Rietveld formulae; multi-channel least squares refinement	Cement or clinker phase analysis	Polymorphism of principal phases
QXRD with selective dissolution	Mass percentages of anhydrous residues after dissolution	Peak height or area comparisons	Cement or clinker phase analysis	Polymorphism of principal phases
Selective dissolution	(*can be followed by chemical, XRD or SEM analysis of residues*)	Mass balance	Mass fractions of major phases in cement or clinker	Compositions of separated phases, etc.
Optical microscopy (reflected light)	Etching techniques for desired phase contrast	Point counting (or image analysis) followed by density correction	Clinker phase analysis	Crystallite size, form, distribution; clinker porosity

Table II, continued

Method	Input data required (optional data)	Computational method used	Principal application	Supplemental data outputs
Optical transmission microscopy	Typical optical properties of clinker phases (Ono method)	Crystal counting	Follow kiln burning conditions	Crystallite size, form, distribution
SEM (backscatter mode)	Gray level standards for main clinker phases	Point counting (or image analysis) followed by density correction	Clinker phase analysis	Crystallite size, form, distribution; clinker porosity
SEM with XRF analyzer (EPMA)	Compositional standards	Point counting (or image analysis) followed by least squares compositional analysis	Clinker phase analysis	Crystallite size, plus elemental composition, form, distribution

Table III. An example in which lime and silica errors reinforce each other

CaO	SiO$_2$	Al$_2$O$_3$	Fe$_2$O$_3$	C$_3$S	C$_2$S	C$_3$A	C$_4$AF
69.2	23.0	4.9	2.9	69.5	13.5	8.2	8.9
68.7	**23.5**	4.9	2.9	63.7	19.3	8.2	8.9
68.2	24.0	4.9	2.9	57.9	25.1	8.2	8.9

the MgO was present in the form of a fifth phase, periclase,[1] but we now know that up to about 2% MgO can dissolve in the four principal clinker phases, as will be discussed below. This also highlights the fact that the Bogue formulae do not even take into account mutual solid solutions of the four principal phases, which can make quite big a difference to the actual quantities found.

Departure from Equilibrium

Even before the above difficulties are considered, we must treat the serious question of departure from equilibrium that occurs because the precursor phases never, in practice, have time to react completely at clinkering temperatures. It is important to remember that the high-temperature clinkering reaction represents the transition from a lower-temperature phase assemblage containing C$_{(free\ lime)}$ + C$_2$S + melt to an assemblage containing C$_3$S + C$_2$S + melt. Thus, the clinkering process can be represented by the reaction:

$$C + C_2S \rightarrow C_3S$$

The degree of completion of this reaction represents the approach to equilibrium, but it is almost never complete under practical conditions, even with LSF values well below 100%, owing to slow diffusion of the reactants. Thus, some free lime almost always remains in clinker as a fifth nonequilibrium phase. But this can easily be treated by using the same Bogue equations to calculate the high-temperature equilibrium phases, but withholding the measured value of the free lime (which must be measured on the actual clinker in question by some specific technique) from the value of C used in the Bogue calculation:

$$C_{(equilibrium\ phases)} = C_{(total)} - C_{(free\ lime,\ measured)}$$

It can be seen that this simple modification already changes the Bogue formulae from being purely predictive into being only partially predictive, since the clinker free lime content must normally be measured on the product itself. Evidently, it depends on the reaction kinetics and the actual conditions used for clinkering.

A second factor that can cause significant divergences from the Bogue-predicted phase composition is that of variable cooling rates. The Bogue formulae themselves do not consider the path of crystallization of the liquid phase on cooling from clinkering temperatures to the solidus. They are, in fact, based entirely on the subsolidus phase equilibria above 1250°C, (because below this temperature C_3S is itself unstable and should decompose into free lime plus C_2S). On the other hand, Lea and Parker's LSF does take into account the true liquid phase composition, which is why it gives an apparently lime-deficient interstitial phase composition. In reality, under typical industrial clinker-cooling conditions, there is not enough time for all of the required free lime to redissolve in the melt to give exactly the aluminate and ferrite compositions used in the Bogue formulae.

Thus, there remain many uncertainties about the exact composition of the crystallized liquid phase in real clinkers, and its composition and reactivity are know to be sensitive to cooling rate as well as to the presence of minor components. For many years it was believed that industrial clinkers contained significant amounts of a glassy phase (undercooled melt), and Lerch provides many interesting data that seem to support this hypothesis.[16] Taylor[4] was doubtful of the existence of a true glass, but Lerch's data certainly do at least clearly demonstrate that the enthalpy of clinker varies with its cooling rate, faster-cooled clinkers generally having a significantly higher enthalpy of solution in a standard acid mixture. Lerch took into account many of the other possible factors in his study, such as belite inversion, free lime variations, and the presence of periclase and of Fe^{II}, but these did not appear to be capable of fully accounting for the observed heat of solution data. Thus, it appears that there must be a term that relates if not to a true glass content, then at least to the state of crystallinity of the quenched clinker. This state could have a higher than expected enthalpy for several reasons other than the presence of glass. Apart from the obvious possibility that the quenched interstitial phase may be very microcrystalline, the quenched clinker may well have an overall phase composition significantly different from that which occurs on slow cooling, as would indeed be expected based on Lea and Parker's work.

In the same vein, Barry and Glasser[17] have recently published a more thorough analysis of the relationships that dictate the phase proportions that would be expected to remain after cooling a clinker at equilibrium at the maximum clinkering temperature. They made use of thermodynamic computer modeling to test several possible scenarios, and their computations clearly demonstrate that the rate of cooling could have a major impact on the final clinker composition. This is essentially because the redissolution of crystalline phases during cooling is relatively slow and thus can probably be ignored in most practical cases. Their calculations show that the alite content in "pure" (C-A-F-S) clinkers should be higher than that predicted by the Bogue formulae, even without considering the effects of solid solution. The total amount of alite would be expected to be higher in a rapidly quenched clinker than in one that is slow cooled; this could account for the significant enthalpy differences observed by Lerch.

Another surprising observation in Lerch's paper was that a rapidly quenched artificial melt phase showed weak diffraction lines in the same locations as those of cubic C_3A, even for a melt phase composed only of calcium, silicon, and iron oxides. This suggests that the composition of the so-called proto-C_3A phase, which is commonly found in rapidly quenched melts,[18] may allow for a very wide range of solid solutions.

An additional kinetic factor that should also be taken into account in real clinkers is the possibility that some of the initial raw materials do not react at all, typically because of their large particle size or high refractoriness. This is commonly the case with quartz, which is hard to grind and thus is often present in rather large, unreactive crystallites that can partially persist through to the clinker. Such material is generally detected in clinker by means of an insoluble residue test (the solid residue remaining after a certain specified acid dissolution procedure). Any such residue must also be excluded from the Bogue calculation, but of course still contributes to the mass of the clinker. This factor also is not predictable simply from the chemistry of the raw mix.

These observations raise a second important question: Should the Bogue calculation be considered a predictive tool or a quality control tool? Given the fact that it cannot, on its own, predict the degree of formation of the desired clinker phases, it is clear that the Bogue calculation alone can never give sufficient information for assessing the chemical and mineralogical quality of a clinker. At the very most, it can perhaps assess the maximum potential quality of a clinker, assuming that the free lime content and insoluble residue of the actual clinker will both be reasonably low. What it is most useful for, however, is designing the raw mix that is to be fed to the kiln to make the clinker in question. In that respect, it is important as a predictive tool for the cement manufacturer, even though it is by no means the only tool necessary to ensure good cement quality.

Modification of the Percentage of the Major Phases Due to Minor Elements

Since all practical kiln feeds will contain many more elements than the five principal ones used in the Bogue calculation, we must expect that the situation in a commercial clinkering process will be much more complicated. There are nine additional elements that are almost always encountered in cement raw materials (including the kiln fuel, which usually contributes significantly to the raw mix composition) in significant quantities. These are Mg, K, Na, Ti, Mn, P, S, H, and C. Of these, the last two are so volatile that they never persist in the clinker in significant quantities. Hydrogen as free water in the kiln feed evaporates readily, and even strongly bound hydroxyl groups in some aluminosilicate raw materials are dehydroxylated well below clinkering temperatures. Carbon as carbonate in the raw materials is also mostly decomposed to give carbon dioxide gas at temperatures below about 1000°C. Carbon and hydrogen from the fuel are usually fully oxidized during combustion. Thus, if any significant levels of H or C are found in clinker, it is almost always due to later contamination (e.g., aeration) and thus is simply a sign of poor sample preparation.

The remaining seven common minor elements all persist in clinker in significant quantities, typically totaling around 5% by clinker mass, and so cannot be ignored in any theoretical prediction of clinker composition. They are also almost always found in their highest common oxidation states, with the possible exception of manganese, which is generally assumed to be present mainly as Mn^{III}. Thus, the common minor oxides in clinker are assumed to be MgO, K_2O, Na_2O, TiO_2, Mn_2O_3, P_2O_5 and SO_3.

Taking into account these seven important minor elements, a typical portland clinker is thus composed of eleven independent oxide components, allowing for a maximum of eleven different condensed phases at equilibrium at the maximum clinkering temperature. Any realistic predictive model must take into account these elements in some way or other, and this is yet another limitation inherent in the use of the four-component Bogue calculation.

In order to deal with this problem, several different approaches have been proposed. Many of these approaches, however, require further analysis of the clinker, and so are not useful as predictive tools. Yamaguchi and Takagi[11] proposed a general approach based on the assumption of specific oxide compositions for the principal clinker phases. The idea is mathematically very simple. If we assume that each phase has a fixed composition in terms of the n component oxides that we wish to use as the basis set (n being four in the basic Bogue calculation), we can write the composition each phase P^i as a vector ($P^i_1 \ldots P^i_n$) in n-dimensional oxide space. The total composition of the clinker in terms of the n oxides is known by chemical analysis and is represented by the vector C ($C_1 \ldots C_n$). The unknowns are then the mass fractions of each of the n possible clinker phases, $M^1 - M^n$, which can also be written as a vector. We thus have the matrix equation $C = PM$, where P is the matrix of phase compositions. This equation is readily inverted to give the value of M if we know M and C. In most cases, we actually assume that there are fewer than n clinker phases, so there is a certain redundancy built in to the analysis, and we can therefore eliminate the least important of the minor oxides from the calculation. If the clinker were truly at equilibrium at a fixed clinkering temperature, and were then quenched infinitely rapidly from that temperature, this analysis should hold for all such clinkers, provided that the total clinker composition were maintained within the n-dimensional space volume limited by the eleven possible equilibrium phases at the clinkering temperature chosen. Evidently, however, the real situation is much more complicated, partly because significant phase changes can occur during cooling, and also partly because not all clinkers occupy the same equilibrium phase space at the clinkering temperature, due to both differences in composition and differences in maximum temperature.

Yamaguchi and Takagi[11] applied these equations to clinkers in which the average compositions of the four principal clinker phases were determined directly on the clinker in question by analytical procedures such as electron microprobe analysis or selective phase dissolution followed by elemental analysis of the residue. But this approach requires careful analyses for each the major phases in each clinker to be analyzed. Hal Taylor was the first to propose a simple and practical compromise,[19] which makes use of average clinker phase compositions established from a wide range of commercial clinkers. It is this method that is currently the most important predictive model beyond the original Bogue calculation. Most importantly, it takes into account the fact that the four major phases in real clinkers do not have the simple compositions assigned to them by Bogue: alite (C_3S), belite

(C$_2$S), aluminate (C$_3$A), and ferrite (C$_4$AF). The oxide formulae that Bogue used were the compositions of pure phases that approximated, in many respects, the crystalline and chemical compositions of the principal phases actually found in commercial clinkers. But they were not based on analyses of commercial clinkers; rather, they were based on the pure four-component oxide system as studied in the laboratory. Taylor's suggested compositions were based on a careful electron microprobe analyses of a large set of industrial clinkers. The average compositions (in mass percentages) of the four principal clinker phases are given in Table IV for all clinkers with alumina ratios (A/F) of between about 1 and 4.

The use of the resulting modified Bogue equations results in an alternative prediction of clinker phase composition that should, in theory, give results that are closer to the true phase composition, so long as the actual free lime and insoluble residue were taken into account in the usual way. Below, we give a comparison of the Taylor modified Bogue calculation with the traditional Bogue calculation for a typical OPC clinker. Imagine that we have a clinker A with the elemental analysis shown in Table V. The traditional Bogue formulae applied to this clinker give 59.8% C$_3$S, 17.5% C$_2$S, 10.3% C$_3$A, and 8.2% C$_4$AF. If we apply Taylor's formulae, we obtain 67.2% alite, 13.8% belite, 9.7% aluminate, and 7.5% ferrite. Note also that if we apply the formulae developed by Cariou and Sorrentino[14] with the assumption that in this case all of the alkalis are soluble, we obtain 67.7% alite, 16.5% belite, 6.6% aluminate, and 4.6% ferrite. This case is perhaps not a good one for Cariou and Sorrentino's formulae because the clinker contains excess sulfate, which no doubt contributes to a lower belite content in the Taylor formulae. (Clearly, some calcium sulfate will be present in this clinker, probably as calcium langbeinite, and this possibility was not explicitly treated by Cariou and Sorrentino.)

However, the mean phase compositions proposed by Taylor represent the averages obtained from many clinkers. It is, therefore, desirable to know the likely variation in the estimation of the percentage content of each phase that is due to the difference between the actual composition of the four principal phases in each clinker analyzed, and the average values used in the Taylor calculation. For example, suppose that in one clinker of interest, the amount of alumina in the alite is actually 2% instead of the 1% assumed in Taylor's formulae. Following the same type of matrix algebra, we can show that the phase percentages change to 72.2% alite, 10.1% belite, 7.1% aluminate, and 8.1% ferrite. Thus, we may also say that a 1%

absolute error in the alumina measurement on the alite of interest (when analyzed by EPMA or equivalent) can lead to a 7% relative error for the alite, 3% for belite, 2% for aluminate, and 1. 5% for the ferrite. However, a 1% absolute error in this case represents a 100% relative error, and we are not likely to be as inaccurate as that! Thus, it seems likely that it is reasonable to use Taylor's average compositions for most clinkers.

For clinkers with very high or very low alumina ratios (e.g., white portland cements [WPC] and sulfate-resisting portland cements [SRPC], respectively), Taylor suggested that somewhat different average phase compositions should be used, as shown in Table VI, although he admitted that there were fewer good data available for such clinkers.

MgO

It should be noted that Taylor's phase calculation takes into account the limited solubility of MgO in the four principal clinker phases. If the total MgO content of the clinker is below 2%, it is assumed to be completely dissolved in the four main clinker phases. Any amount above 2% is assumed to persist as the separate crystalline phase periclase. Given that the MgO contents of commercial clinkers can in some cases run as high as 5%, this effect is very important. As Taylor points out, MgO has little effect on the calculated content of aluminate or ferrite, but a significant effect on the amount of alite and belite. It is further noted that the Taylor calculation generally gives a higher amount of alite and a lower amount of belite than the simple Bogue calculation, a result that is in better accord with direct methods of clinker analysis, such as microscopy or quantitative X-ray diffraction.

Mg^{2+} substitutes for Ca^{2+} in alite and belite. The proportion of this substituent Mg that enters alite usually lies between 0.74 and 0.65,[11,20] leaving only 0.25–0.35 in the belite. This leads to an increase in the amount of alite relative to belite in the Taylor calculation relative to the Bogue calculation. The maximum amount of MgO in the solid solutions are, reportedly,[21,22] 0.98–1.60% in alite; 0.40–0.52% in belite; 0.80–1. 97% in the aluminate; and 3.90–4.36% in the ferrite. Thus, one would also expect MgO to increase the amount of ferrite phase, as is indeed often observed.

If we assume that these values don't depend on the clinkering temperature or the cooling rate, it is possible to incorporate them into modified lime saturation factors[21] as follows:

$$LSF = 100 \{(CaO + 0.75\,MgO) / (2.80\,SiO_2 + 1.18\,Al_2O_3$$

Table IV. Average compositions (mass%) of the four principal clinker phases, obtained from EPMA on a large set of industrial clinkers[19]

	Na$_2$O	MgO	Al$_2$O$_3$	SiO$_2$	P$_2$O$_5$	SO$_3$	K$_2$O	CaO	TiO$_2$	Mn$_2$O$_3$	Fe$_2$O$_3$
Alite	0.1	1.1	1.0	25.2	0.2	0.0	0.1	71.6	0.0	0.0	0.7
Belite	0.1	0.5	2.1	31.5	0.2	0.1	0.9	63.5	0.2	0.0	0.9
Aluminate	1.0	1.4	31.3	3.7	0.0	0.0	0.7	56.6	0.2	0.0	5.1
Ferrite	0.1	3.0	21.9	3.6	0.0	0.0	0.2	47.5	1.6	0.7	21.4

Table V. Example: Elemental analysis (mass%) of a typical clinker

	Na$_2$O	MgO	Al$_2$O$_3$	SiO$_2$	P$_2$O$_5$	SO$_3$	K$_2$O	CaO	TiO$_2$	Mn$_2$O$_3$	Fe$_2$O$_3$
Clinker A	0.07	1.26	5.55	20.92	0.04	1.40	0.42	65.81	0.26	0.05	2.65

Table VI. Taylor's proposed compositions (mass%) for WPC and SRPC clinker interstitial phases[19]

Phase	Na$_2$O	MgO	Al$_2$O$_3$	SiO$_2$	P$_2$O$_5$	SO$_3$	K$_2$O	CaO	TiO$_2$	Mn$_2$O$_3$	Fe$_2$O$_3$
Aluminate (WPC)	0.4	1.0	33.8	4.6	0.0	0.0	0.5	58.1	0.6	0.0	1.0
Ferrite (SRPC)	0.1	2.8	15.2	3.5	0.0	0.0	0.2	46.0	1.7	0.7	29.8

Table VII. Typical composition (mass%) of an alkali-aluminate, usually orthorhombic[19]

Phase	Na$_2$O	MgO	Al$_2$O$_3$	SiO$_2$	P$_2$O$_5$	SO$_3$	K$_2$O	CaO	TiO$_2$	Mn$_2$O$_3$	Fe$_2$O$_3$
Alakai-aluminate	0.6	1.2	28.9	4.3	0.0	0.0	4.0	53.9	0.5	0.0	6.6

$$+ 0.65\,Fe_2O_3)\}, \text{ for MgO} < 2\%$$

$$LSF = 100\,\{(CaO + 1.50\,MgO) / (2.80\,SiO_2 + 1.18\,Al_2O_3$$
$$+ 0.\,65\,Fe_2O_3)\}, \text{ for MgO} > 2\%$$

However, given the complexity of the substitutions actually involved and the resulting phase diagram, it is doubtful that these modified LSF equations are very precise, and their usefulness is probably simply a question of local convenience.

Alkalis and Sulfates

Taylor also pointed out that it was essential to run a separate calculation for the sulfate-rich phases in the clinker, which were not taken into account at all by the original Bogue calculation but, as is now well-recognized, play an important role in clinker quality. To take care of this problem, Taylor incorporated a model based on the work of Pollitt and Brown.[23] Essentially, their work showed that the sulfates in clinker form in many cases were a separate liquid phase that is immiscible (even at clinkering temperatures) with the principal clinker liquid phase (C-A-F-S-M). The amount of this sulfate-rich liquid phase depends not only on the total percentage of clinker sulfate (SO_3) but also on the ratio of sulfate/total alkali (R), and also on the ratio of potassium over sodium (k) (all expressed as oxides by mass).

In fact, it is observed that the quenched sulfate melt phase persists in clinker as soluble alkali sulfates; that is, it is readily soluble in water and can be measured qualitatively by a water extraction of clinker under suitable circumstance, followed by an analysis of the alkali metals in solution (soluble alkalis). However, this approach is no longer predictive, because it requires yet another analytical test on the clinker. Moreover, only the alkali sulfates can readily be measured in this way, and not the soluble calcium sulfate that may also exist in the clinker, either as anhydrite $(C\bar{S})$ or as calcium langbeinite $(KC_2\bar{S}_3)$, both being possible depending on the values of R and k. These two phases are of increasing interest nowadays, as they occur in clinkers with high values of R, and such clinkers are becoming increasingly abundant owing to the increased use of high-sulfur fuels. The presence of significant quantities of calcium langbeinite in clinker can have very significant effect on the early age properties of the cement, on its storage stability, and especially on concrete rheology.

Taylor summarized Pollitt and Brown's results as follows:

- $\bar{S}_s$ is defined as soluble SO_3 (effectively, all the readily soluble clinker sulfate). We use the notation $\bar{S}$ to represent the total clinker SO_3.

- K_s is defined as sulfate K_2O (effectively equivalent to soluble K_2O).
- N_s is defined as sulfate Na_2O (effectively equivalent to soluble Na_2O).
- C_s is defined as sulfate CaO.

We also define three separate factors based on the total clinker composition:

$$R = \bar{S} / (N + K) \ \{= [SO_3 / (Na_2O + K_2O)]\}$$

$$r = (K + N) / (1.12\,K + 0.56\,N)$$

$$k = (K / N)$$

We then have different formulae for different values of R and k, as follows:

for $R \le 0.8$: $\bar{S}_s = \bar{S}$

if $k < 3.67$: $K_s = 1.12\,RK$; $N_s = 0.56\,RN$

if $k \ge 3.67$: $K_s = 1.12r\,RK$; $N_s = 0.56r\,RN$

for $0.8 < R \le 2$: $\bar{S}_s = [1.0 - 0.25\,(R - 0.8)]\,\bar{S}$

for $R > 2$: $\bar{S}_s = 0.7\,\bar{S}$

and in both cases (i.e., for $R > 0.8$):

if $k < 3.67$: $K_s = 0.9\,K$; $N_s = 0.45\,N$

if $k \ge 3.67$: $K_s = 0.9r\,K$; $N_s = 0.45r\,N$

Finally, to balance the excess "soluble" sulfate with calcium, we have:

$$C_s = \bar{S}_s - (K_s + N_s)$$

In the event that the value of R is low, a significant fraction of the clinker alkalis will dissolve in the four principal clinker phases, and, in particular, in the aluminate phase. This can give rise to the existence of two different aluminate phases: the standard (cubic) C_3A, and another, richer in alkalis, which is generally orthorhombic and whose composition also depends on the potassium/sodium ratio *(k)*. Taylor's average composition for this phase is shown in Table VII. Although Taylor's approach to the treatment of the clinker sulfate phases and soluble alkalis is reasonably comprehensive, it is based largely on the experimental data of Pollitt and Brown.[23] Recent studies, although not calling into question the general tendencies, have added some further complications. For example, Glasser[24] observes that the alkali sulfate phases found in many industrial clinkers are frequently not pure sulfates, but can instead contain a significant fraction of carbonate ion replacing sulfate. When this ion goes into solution, it will probably behave essentially as hydroxide, since the carbonate will be rapidly precipitated by lime. However, this reaction could influence the early hydration kinetics of the cement. This possibility highlights the difficulty of accurately measuring the alkali sulfate content of a clinker, since the standard soluble alkali method (itself a form of selective dissolution) can be significantly influenced by the way in which the experiment is carried out, and is surely not truly quantitative. The sulfate can also be precipitated rapidly (as ettringite), so it is not possible to obtain an accurate sulfate-alkali balance for the water soluble fraction. All of these complications are compounded by the fact that clinker alkali sulfate contents are usually 1% or less, so analytical errors can be relatively large, and the phases are almost impossible to analyze accurately either by QXRD or by microscopic point counting.

An alternative approach of some value is to use the method of Cariou and Sorrentino,[14] that is, by using an EPMA, or a well-calibrated SEM with XRF detector, to estimate the alkali and sulfur contents of the major crystalline phases of the clinker, and then calculate the alkali and calcium sulfate fractions by difference. This approach avoids the selective dissolution techniques used by Pollitt and Brown, but its accuracy is also limited, especially for sodium.

Phosphates

Significant levels of phosphorus in portland cement clinker can come from normal cement raw materials, as is frequently the case in parts of Africa (e.g., Morocco, Uganda, etc.) where limestone tends to be relatively rich in calcium phosphate. Alternatively, it may come from the use of industrial by-products, such as metallurgical slags, animal meals, or sewage sludges, either as raw feed additives or as kiln fuels. Whatever the source of phosphorus, it is generally found only in its pentavalent oxidation state (i.e., as phosphate) in clinker. The orthophosphate ion is isomorphous with the orthosilicate ion as found in alite or belite, but has one less unit of negative charge. Thus, quite high degrees of phosphate substitution can easily occur in calcium silicates, but with the condition that one less unit of positive charge is needed to balance it in the lattice. This should lead either to a change in phase composition and symmetry, or to preferential substitution of an alkali metal ion of similar radius (e.g., sodium) for calcium along with the substitution of an orthophosphate ion for orthosilicate. It is apparently not yet known how combined substitutions of sodium compounds and calcium phosphates interact. However, the effect of another secondary substituent, the fluoride ion (which is also found in significant amounts in some phosphatic raw materials), has been shown to be beneficial for alite formation in the presence of phosphate,[25] although the mechanism is not very clearly understood. This observation highlights the complexity of sets of impurities that can, and frequently do, have a tendency to act in a concerted and thus highly nonlinear manner. (Note that alkali and sulfate are good examples of another pair of minor components that act in a concerted manner, although in this case the mechanism is well understood.)

In practice, it is observed that the phosphate in clinker is present mainly in the belite phase, although a significant amount (up to 2%, according to some authors) can also be found in alite. Furthermore, whenever the phosphate content of the clinker is high, the clinker tends to contain much more free lime than would be predicted from the Bogue formulae if phosphate is not taken into account. This led many researchers to conclude that phosphate stabilizes belite plus free lime relative to alite at clinkering temperatures. However, this is an oversimplification, as was shown by the work of Nurse's team at the Building Research Establishment (to be discussed below).

If one uses the method of Lea and Parker for preferentially combining the oxides, one assumes that Fe_2O_3 first gives C_4AF, the remaining Al_2O_3 gives C_3A, and the CaO, SiO_2, and P_2O_5 remain to form belite, alite, and/or free lime. The system CaO-SiO_2-P_2O_5 was first studied by Barrett and McCaughey.[26] They found no alite in this system at any temperature (liquidus or solidus) but simply two primary phase assemblages:

1. CaO, β-C_2S, and nagelschmidtite (to which they assigned the formula C_7PS_2).
2. β-C_2S, C_3S_2, and nagelschmidtite.

In a later study, in the diagram C_2S-C_3P, Lea and Nurse[27] identified a compound of formula C_7PS_2 (called "A") that was structurally different from nagelschmidtite. Trömel[28] identifies a phase structurally similar to nagelschmidtite. Finally, Nurse, Welch, and Gutt proposed to call the whole range of solid solution between α-C_2S and C_3P by the name "nagelschmidtite."

Further work by Nurse[29] showed that a particular composition situated on the line C_2S-nagelschmidtite contained alite, a solid solution of C_2S/C_3P, and 10% free lime, and consequently proved the existence of C_3S and C_3S solid solution in the solidus of the system CaO-SiO_2-P_2O_5.

The composition of the solid solution between C_2S, CaO, and P_2O_5 that exists in equilibrium with C_3S and CaO at 1400°C (referred to as PSS) is approximately 7% P_2O_5, 26.5% SiO_2, and 66.5% CaO. Nurse[29] established the following set of clinker phase equations based on the assumption that this PSS phase would persist in clinker regardless of whether or not excess free lime was present:

(i) $C_3S = 4.07\,CaO - 7.60\,SiO_2 - 6.72\,Al_2O_3 - 1.43\,Fe_2O_3 - 9.9\,P_2O_5$

(ii) $C_2S = 8.60\,SiO_2 - 3.07\,CaO + 5.10\,Al_2O_3 + 1.08\,Fe_2O_3 - 3.4\,P_2O_5$

(iii) $C_3A = 2.65\,Al_2O_3 - 1.69\,Fe_2O_3$

(iv) $C_4AF = 3.04\,Fe_2O_3$

(v) $PSS = 14.3\,P_2O_5$

In some respects the PSS phase can simply be treated as additional belite, in which case, by adding PSS to C_2S in the above formulae, we see that we obtain the standard Bogue equations to which 9.9 P_2O_5 has been subtracted from the alite and added to the belite.

Gutt and Smith also applied Nurse's formulae to the case of Ugandan raw materials.[30] To calculate the lime required to saturate this system, Nurse proceeded as follows:

1. Calculate the CaO necessary to form C_3A and C_4AF (= 1.65 Al_2O_3 + 0.36 Fe_2O_3).
2. Subtract this from the total CaO to give the remaining CaO (CaOr).
3. The composition remaining is therefore CaOr + SiO_2 + P_2O_5.
4. At lime saturation, the SiO_2 content of this mixture must be 26.5%.

5. Therefore,

$$SiO_2 / (CaOr + SiO_2 + P_2O_5) = 0.265$$

Solving the above set of equations, we obtain:

$$CaO_{(sat)} = 2.77\,SiO_2 + 1.65\,Al_2O_3 + 0.36\,Fe_2O_3 - P_2O_5$$

Alternatively, we can take Nurse's equation (ii) above and set it to zero, on the assumption that, at lime saturation, we will have no belite (only PSS). Doing this results in another equation that is fairly close to the one derived above:

$$CaO_{(sat)} = 2.63\,SiO_2 + 1.66\,Al_2O_3 + 0.35\,Fe_2O_3 - 1.11\,P_2O_5$$

Whichever of these approaches is chosen, it is clear that addition of phosphate reduces the total amount of lime that can be combined in clinker and sets an upper limit to the possible alite content that can be obtained, because the PSS phase cannot be converted to alite. Nevertheless, even at apparently high phosphate contents (say, 1%), clinkers can theoretically be made with quite normal alite contents. This says nothing at all, however, about their hydraulic reactivity, which can be greatly reduced.

The phase diagram for the C-S-P system in the region of interest was put together in stages by Nurse's group at the BRE. Nurse et al.[31] first completed the join C_2S-C_3P, Welch and Gutt[32] added the join CaO-C_2P, and Gutt[33] completed the system C_2S-C_3P-CaO. He found a phosphate and lime modified tricalcium silicate (C_3S') containing 0.5% P_2O_5 and defined a limit of the occurrence of C_3S at 1500°C (PSS' of 86.9% C_2S, 13.1% C_3P). The phase assemblages in the relevant part of this three-component system at 1500°C are:

1. C_3S and solid solution of C_2S-PSS'.
2. Solid solution C_3S-C_3S',PSS', and CaO.
3. Solid solution C_3S-C_3S' andPSS'.
4. Solid solution C_3S-C_3S' andCaO.

$(C_3S)_s$ disappears when P_2O_5 is greater than the line CaO-PSS (Fig. 2).

Other Minor Components

In addition to the range of compositions that exist in common clinkers, these days we are forced to also consider cases in which certain of the

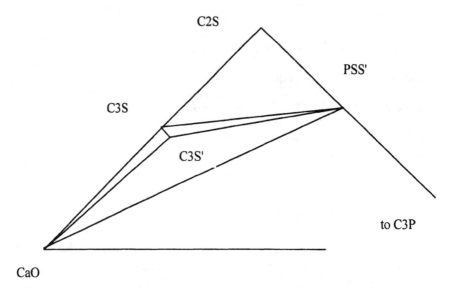

Figure 2.

minor components (e.g., P) are present in relatively high concentrations that perturb significantly the high-temperature equilibria. We may also have to consider significant concentrations of other elements, such as Zn, Cr, Ni, and Cl, that may be introduced from certain industrial by-products used as fuels or raw materials.

Conclusions: Prediction vs. Quality Control

As stated earlier, the principal value of the original Bogue calculation is its predictive value for the cement manufacturer in terms of raw mix preparation. That is no doubt why its use has persisted so long in the industry. Every cement plant is forced to deal with the problem of selecting and blending its local raw materials in order to make a consistent clinker. Even with only four main components to worry about, the need to obtain a consistent raw mix that can, in principle, give a consistent quality clinker is quite a significant operational challenge. From that viewpoint, Taylor's calculation can be seen as a true advance, in the sense that it is predictive and that it can be restricted to a consideration of the four main clinker phases. However, it does nevertheless add complexity for the cement manufacturer, and this complexity is not always welcome, since the manufacturer frequently does not have the option to change many of the raw material variables. It is for this reason that most still use the classic Bogue formulae to design their kiln feeds.

It is evident that neither the Taylor method nor the original Bogue method can ensure the actual quality of the clinker produced by any given plant, because that depends also on the way in which the kiln is operated. Furthermore, the quality of the cement produced by the plant depends on other downstream factors, such as the operation of the grinding mills and the amount and type of calcium sulfates or other additions used in the final grinding and blending steps. As a result of this complexity, the best that a cement producer can usually do is to use some form of statistical quality control based on a correlation between the composition of the clinker produced and the results of standard mortar or concrete tests on the resulting cements. As input to the statistical analysis, the producer can use the chemical composition of the clinker expressed in any format that he finds convenient, but the use of the Bogue composition is generally accepted because the use of more than four independent components usually becomes very unwieldy. Since the Bogue equations are simply based on a linear combination of the concentrations of C, S, A, and F in the clinker, the manufacturer could equally well use these four oxides directly in the statistical analysis if it is in the form of a multiple linear regression, which is usually the case.

This approach changes somewhat if the principal objective of the cement manufacturer is to control the operation of his kiln, rather than to design his kiln feed. In that case, the direct observation of clinker can be a very useful quality control tool. Direct phase analysis by the Rietveld method is also useful here, as are the more conventional techniques of free lime and soluble alkali determination. In fact, it seems that the majority of modern cement quality issues are related to the state of the sulfate phases, both in the clinker and in the cement. This cannot be addressed simply by a calculation, but should be considered as the primary tool for cement quality control after the major clinker phases and free lime content are held constant by the more standard procedures, for which the traditional Bogue calculation is probably largely sufficient.

References

1. R. H. Bogue, "Calculations of Compounds in Portland Cement," *Ind. Eng. Chem.* (Analytical ed.), **1** [4] 192 (1929).
2. F. M. Lea and T. W. Parker, "Investigations on a Portion of the Quaternary System

$CaO-Al_2O_3-SiO_2-Fe_2O_3$," *Phil. Trans. R. Soc. Lond.*, **A234**, 1 (1934).

3. R. H. Bogue, "Compounds in Portland Cement," *Ind. Eng. Chem.*, **27**, 1312 (1935).

4. H. F. W. Taylor, *Cement Chemistry*, 2nd ed. Thomas Telford, 1997.

5. I. Quirina, R. Gutzmer, and Y. Ballim, "Phase Composition and Quantitative X-Ray Powder Diffraction Analysis of Portland Cement and Clinker"; pp 1–49 in *Materials Science of Concrete VI*, edited by S. Mindess and J. Skalny. American Ceramic Society, Westerville, Ohio, 2001.

6. S. Brunauer, L. E. Copeland, et al., "Quantitative Determination of the Four Major Phases in Portland Cement by X-Ray Analysis," *Proc. ASTM*, **59**, 1091 (1959).

7. W. Klemm and J. Skalny, "Selective Dissolution of Clinker Mineral and Its Application"; in *Proc. Siliconf.* Budapest, 1977.

8. H. M. Rietveld, "Line Profiles of Neutron Powder Diffraction Peaks for Structure Refinement," *Acta Cryst.*, **22**, 151–152 (1967).

9. M. P. Javellana and I. Jawed, "Extraction of Free Lime in Portland Cement and Clinker by Ethylene Glycol," *Cem. Conc. Res.*, **12**, 399–403 (1982).

10. F. J. Tang, PCA/CTL Research and Development Bulletin RD 105T. 1992.

11. G. Yamaguchi and S. Takagi, "The Analysis of Portland Cement Clinker,"; 181–225 in *Proceedings of the 5th International Symposium on the Chemistry of Cement*, vol. 1. Tokyo, 1968.

12. J. P. Skalny, J. E. Mander, et al., "SEM Study of Partially Dissolved Clinkers," *Cem. Conc. Res.*, **5**, 119–128 (1975).

13. D. Campbell, *Microscopical Examination and Interpretation of PC Clinker.* Portland Cement Association, Skokie, Illinois, 1999.

14. B. Cariou and F. Sorrentino, "Industrial Application of Quantitative Study of Portland Cement Clinker through Reflected Light Microscopy"; pp. 277–284 in *ICMA Proceedings.* San Antonio, 1988.

15. Y. Ono, "Microscopical Estimation of Burning Conditions and Quality of Clinker"; pp. 206–211 in *Proc. 7th ICCC*, vol. 2, theme 1. Paris, 1980.

16. W. Lerch, "Approximate Glass Content of Commercial Portland Cement Clinker," *J. Res. Natl. Bur. Stds.*, **20**, 77–81 (1938).

17. T. I. Barry and F. P. Glasser, "Calculations of Portland Cement Clinkering Reactions," *Adv. Cem. Res.*, **12** [1] 19–28 (2000).

18. R. M. H. Banda and F. P. Glasser, "Crystallisation of the Molten Phase in PC Clinker," *Cem. Conc. Res.*, **8**, 665–670 (1978).

19. H. F. W. Taylor, "Modification of the Bogue Calculation," *Adv. Cem. Res.*, **2** [6] 73–77 (1989).

20. M. Kristmann, "Portland Cement Clinker Mineralogical and Chemical Investigations P1 and P2," *Cem. Conc. Res.*, **7**, 649 (1977); **8**, 93 (1978).

21. M. V. Rangarao "Effect of Minor Components on Formation and Properties of Portland Cement Clinker," *Cement Bombay*, **10**, 6–13 (1977).

22. H. Uchikawa and S. Hanehara, *Recycling of Waste as an Alternative Raw Material and Fuel in Cement Recycling.* Onoda Cement Company, Tokyo, 1995.

23. H. W. W. Pollitt and A. W. Brown, "The Distribution of Alkalis in Portland Cement Clinker"; pp. 322–333 in *Proceedings of the 5th International Symposium on the Chemistry of Cement*, vol. 1. Tokyo, 1968.

24. F. P. Glasser, personal communication.

25. W. Gutt; p. 93 in *Proc. 5th ICCC,* vol 1. Tokyo, 1969.

26. R. L. Barrett and W. J. Mc Caughey, *Amer. Min.,* **27**, 680 (1942).

27. F. M. Lea and R. W. Nurse, Ministry of Supply Monograph 11/108. London HMSO, London, 1951.

28. G. Trömel, H.-J. Harkort, and W. Hotop, "The System CaO-P_2O_5-SiO_2," *Z. Anorg. Allgem,* **256**, 253–272 (1948).

29. R. W. Nurse, "The Effect of Phosphate on the Constitution and Hardening of Portland Cement," *J. Appl. Chem.,* **2** [12] 708 (1952).

30. W. Gutt and M. A. Smith, "Studies of Phosphatic Portland Cement"; in *Proc. 6th ICCC,* Moscow, 1974; BRE current paper CP95/74 (1974).

31. R. W. Nurse, J. H. Welch, and W. Gutt "High Temperature Phase Equilibria in the System C_2S/C_3P," *J. Chem. Soc.,* **220**, 1077 (1959).

32. J. H. Welch and W. Gutt, "High Temperature Studies of the System Calcium Oxide Phosphorus Pentoxide," *J. Chem. Soc.,* 874 (1961).

33. W. Gutt, "High Temperature Phase Equilibria in the System C_2S-C_3P-CaO," *Nature,* **197**, 142 (1963).

Thermodynamics of Cement Hydration

F. P. Glasser

Introduction

Thermodynamics provides a mathematically exact and self-consistent framework for describing the physical universe. It includes all classes of chemical reactions, homogeneous as well as heterogeneous. Of course, we usually want to be less ambitious in our compass than describing the entire universe and wish only to describe selected portions. Portions of the universe that can be isolated for separate study are termed a "system": this important concept and the application of thermodynamics to cement hydration deserve general comment.

Mixtures of cement and water comprise a heterogeneous system, and it is therefore possible in principle to apply thermodynamic concepts to hydration reactions. The energy changes occurring in the course of hydrating cement are superficially apparent: experience tells us that hydrating cement evolves much heat, termed "enthalpy." Indeed, when large masses of concrete are made it is essential to plan for release of this heat so as not to entail an excessive temperature rise of the hydrating mass. However, a general conclusion arising from this experience is that while chemical equations alone can describe mass and mineralogical balances, analysis of the driving forces for reaction also involves energy change; therefore a comprehensive account of cement hydration also involves the application of thermodynamics. As will be shown, application of thermodynamics also enables deeper understanding to be achieved of the features, events, and processes occurring in the course of reaction.

The theoretical foundations of thermodynamics have been complete for over a century and are thus well understood. Application of thermodynamics to cement hydration also has a long history. Mchedlov-Petrossyan and colleagues did much valuable work, although much of this work only became accessible outside Russia well after its completion, when it was translated first into German and subsequently into English.[1] The English version contains an exposition of thermodynamic methods and examples of applications showing calculations as well as an appendix of thermodynamic data for cement substances. The quality and documentation of its thermodynamic database is variable; moreover the authors were unable to take advantage of modern computational methods and the resulting scope of the calculations is simplistic by modern standards. Nevertheless, the book stands as a landmark.

This chapter is not intended to be a treatise on thermodynamics; it is a starting point. Accordingly a minimum of formal exposition and definition is presented. Many standard textbooks are available that give, if required, proofs and derivations of equations. After a brief introduction to cements, the underlying thermodynamic concepts are presented and developed using illustrative examples.

The Nature of Cement and Its Hydration

By "cement" is generally meant portland cement. Portland cement is a manufactured product made by high-temperature processing of raw materials, including calcium carbonate as an important component of the raw mix. Complex legal and technical specifications exist governing the composition of portland cement and, while these may vary from one country to another, most specifications are broadly similar. These specifications were reviewed in Ref. 2 but are being revised, often at short intervals.

Finished cement material is supplied to the customer as a fine, dry powder. Mineralogically, this cement powder is not homogeneous but contains four main minerals. In order of decreasing abundance, these are alite, close to tricalcium silicate, Ca_3SiO_5, in composition; belite, close to Ca_2SiO_4 in composition; ferrite, close to $Ca_2(Fe,Al)_2O_5$ in composition, but with varying ratios of iron and aluminium; and tricalcium aluminate, close to $Ca_3Al_2O_6$ in composition. During cooling of the calcined cement clinker, alkali sulfates frequently condense out of the kiln atmosphere onto solids, initially as a melt or in solid form, with formation of alkali and calcium-alkali sulfates. Additionally the clinker may contain a little unreacted free lime, CaO. In the course of grinding calcined cement to a fine powder, several weight percent of a calcium sulfate source (gypsum, anhydrite, hemihydrate, or mixtures thereof) is normally added. Thus, depending on what minor phases are excluded as unimportant, the cement powder delivered to the customer may contain six phases; five if free lime is excluded and calcium and alkali sulfates are counted as two phases. To initiate set and strength gain, cement powder is reacted with water, adding yet another component. Since the complexity of a given system increases factorially with the number of components, it is not surprising that cement hydration is complex and as yet not fully understood.

Many permissive changes to cement specifications have recently been made. Typically, these changes allow the introduction of up to 5–10% of miscellaneous reactive admixtures such as powdered limestone, slag, fly ash, and so on, while still permitting the resulting product to be described as portland cement. These additions complicate the chemistry of the hydration reactions although, fortunately, the level of addition of admixtures permitted in portland cement is low. Of course blended cements containing relatively large amounts of slag, fly ash, and other reactive supplementary cementing materials have been used for centuries. Despite their long history of use, reactions and reaction kinetics occurring in these blended cement systems are complex and usually incompletely understood. Moreover, iron blast-furnace slag introduces substantial quantities of another component, MgO, to the system, thus further increasing the chemical complexity of the matrix inasmuch as the saturation limit of magnesium in the main hydrate phases is exceeded and new phases containing essential magnesium appear.

At or near the point of construction, portland cement is mixed with mineral aggregate. Mixtures with fine aggregate are termed "mortar" while mixtures of cement with both fine and coarse aggregate are termed "concrete." The aggregate dilutes the heat of hydration of cement, helps control dimensional changes occurring in the course of hardening, and lowers cost. The aggregate is not normally reactive with cement on ordinary time scales (10–100 years), although important exceptions to this generalization may occur. Thus mineral aggregates may, under some circumstances, comprise a reactive portion of the system; in these circumstances they alter chemical balances. While reactions involving mineral aggregates can in principle be treated by thermodynamics, they are excluded from this presentation, which will focus on the hydration of cement and blended cement portions, termed "pastes," with brief mention of the performance of cement pastes in service conditions.

The evolution of heat during the course of cement hydration is, as noted, readily apparent. However the rate of heat evolution as a function of time is complex. Figure 1 shows the pattern of heat evolution from a modern portland cement, as determined by isothermal calorimetry. It is immediately apparent that a complex series of events must occur during the early stages of hydration in order to achieve the observed pattern of heat evolution. While we may not be able to deconvolute the underlying events without recourse to additional data, it is apparent that kinetic controls must modify the sequence and course of reactions, especially during the early stages of hydration.

Studies of the thermodynamics and kinetics of cement hydration have tended in recent decades to move apart. This is unfortunate because thermodynamic principles underlie kinetic theory and in practical situations, of which cement hydration is one, the two approaches are best joined: they are not independent of each other.

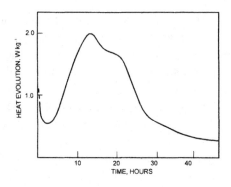

Figure 1. Rate of evolution from a modern portland cement during early hydration. The water/cement weight ratio is ~0.5.

Thermodynamics has also been said — mainly in oral argument, as this author has never found anyone with the courage of conviction to advance the argument in print — to be inapplicable to cements because of the pronounced metastability of certain of the hydrate phases occurring in cement systems. This argument is quite untrue; examples of metastable equilibria and their thermodynamic treatment will be given subsequently. Thus, while thermodynamics is most often used to calculate equilibrium, nonequilibria are also amenable to treatment, especially when the nonequilibrium states are persistent and accessible to measurement.

Those who investigate cements, whether as practical engineers or research scientists, routinely, albeit perhaps unknowingly, employ the language and concepts of thermodynamics; for example, the concept of a "system" is widely used in planning laboratory experiments and in interpreting data. Thus we may not always explicitly use formal thermodynamic equations or calculations but we do borrow thermodynamic concepts to assist in the design of experiments, analysis of data, and the interpretation of both laboratory and field evidence in respect of concrete performance.

The cement and concrete industry, as well as the research community, have recently come under pressure to improve the long-term performance of concrete while at the same time reducing costs, energy inputs, and environmental impacts of production and use. It is apparent to this author that purely empirical solutions, albeit with a veneer of sophisticated instrumentation, statistical design, and so on, permit only very slow progress to be made. Moreover, the results obtained are often incapable of independent validation and cannot readily be extended or extrapolated without the envelope of test conditions. So, a return to fundamentals is occurring. Where do we find validated methodologies that could potentially be applied? Thermodynamics emerges as a strong contender. It has its roots in physical reality and, moreover, because of recent advances in computational methods it is now much more feasible to undertake the necessary thermodynamic calculations associated with chemically complex systems but without the traditional large labor input.

The Thermodynamic Approach: Basic Concepts

Systems

Conventional thermodynamic treatments assume that mass and energy are conserved; that is, neither matter nor energy can be created or destroyed in the course of ordinary chemical reactions. However, energy can be — and often is — redistributed. Strictly, a system is closed with respect to transport of both mass and energy; such a system is termed "adiabatic." Despite the restrictive conditions imposed by this definition, hydrating cement can closely approximate an adiabatic system. For example, if a large mass of concrete is formed, it often closely approximates closed system behavior with respect to both mass and energy, especially in the first few days, when thermal losses or gains by conduction are often negligible. On the other hand, some geometries (e.g., thin slabs) may permit heat to escape readily; although the chemical composition of the reacting mass may not change in the course of hardening and strength gain, energy exchanges occur. To treat these types of commonly encountered situations, we may need to relax somewhat the restrictions imposed by maintaining strictly adiabatic conditions. Thus, for example, a thin slab may comprise a system that is closed with respect to transport of matter but not necessarily of energy; such a system is still described as closed but with the (often unstated) premise that this refers only to gain or loss of matter. On the other hand, to pursue the example of a thin slab with a free surface, limited chemical exchanges may occur; the surface of our hypothetical slab may lose moisture and also react with atmospheric carbon dioxide in the course of early hydration. In this case, the system becomes open with respect to both transport of energy and mass. The system concept is diluted but retains value because mass transport is restricted to relatively few species: in this example, of water and CO_2.

The complexity of treating different types of systems is such that totally closed systems are generally easiest to calculate; systems open to transport of energy, but not of matter, are next in order of increasing complexity, while partially open systems comprise the most complex set of conditions for thermodynamic treatment. Note that the definition of "system" implies some restriction: to be a system, it must be closed, at least in part. Consequently, any restrictions that can be placed on the degree of openness of the system usefully reduce the complexity of the calculations necessary to define its state or condition, or of changes to its state or condition. The nature of these restrictions is important to the limit the range of application of calculations that might be undertaken, and a very important part of applying thermodynamics lies in defining the system, including the extent to which calculations can be restrained, and in ensuring that comparisons between different systems take account of the nature of the restrictions, if any, to ensure that comparisons are valid.

Laboratory and field observations are important in defining change; they help define and rank in importance the nature of cement aging processes and the important mass exchanges. Unfortunately, closed system behavior is rare: observed changes may include environmentally conditioned reactions. However, field studies have distinct limitations in elucidating the origin of change. Some of the limitations are obvious: for example, conditions may fluctuate in the course of exposure and cannot be fixed with respect to temperature as well as other variables (e.g., moisture state). Other limitations are less obvious: for example, while mass gains can often be inferred from the nature of the mineralogical changes occurring in affected cement paste, mass losses are often not directly revealed. Furthermore, both mass loss and gain are difficult to quantify. Field studies do, however, provide a snapshot in time and usefully disclose how kinetic factors interact with equilibria, albeit in a qualitative manner.

Phases

Cement solids are not homogeneous but consist of mixtures of solids, termed "phases." A phase is a region having a constant composition, or a composition that is variable over known limits, and that can in principle be mechanically separable from other phase regions. Although it may be impractical to demonstrate physical separation of the constituent phases of cement pastes, other operational definitions of phases can be substituted. For example, we normally identify the phases present in the hydrated paste on the basis of morphology, microchemistry, X-ray powder diffraction pattern, infrared or Raman spectroscopy, NMR spectra, and so on. Moreover, the presence of phase boundaries is often revealed by electron and optical microscope imaging coupled with chemical-analytical facilities. While an individual technique may not always be diagnostic for specific phases, a collective interpretation of the identity and number of phases usually emerges from comprehensive multitechnique studies that also take account of the findings from laboratory studies. Hence the phases present in cement pastes can be characterized.

It should be noted that the definition of a phase is in part scale dependent. For example, crystalline phases are characterized by having long-range ordering of their constituent atoms, ions, or molecules; these characteristic units, or crystallites, are typically larger than 100 nm. This is because the identity of a crystalline phase usually breaks down at <100 nm (this limit is approximate) as an increasing number of atoms or ions are at surfaces and, as a result, differ in energetics and bonding from those in bulk matter. We shall in general treat only particles ≥100 nm, approximately, such that classical thermodynamics apply, but make reference to size and scale effects especially in respect of C-S-H.

Degrees of Freedom, or Variance

The state or condition of a system is a function of composition, as has been shown, but is also affected by physicochemical parameters. The most commonly used parameters linking chemistry and mineralogy to the real world are temperature and pressure. The variance, or number of degrees of freedom, can be linked to the number of phases and components by the Gibbs phase rule.[3] The state or condition of a system is fixed by defining the numerical value of the degrees of freedom, perhaps by fixing pressure, temperature, or composition.

Phase Rule

Treatment of heterogeneous systems, consisting of mixtures of solids, liquid, and gas, benefit from the application of special rules limiting their constitution. The Gibbs phase rule relates the maximum number of phases occurring at equilibrium to the components and the degrees of freedom of a system at equilibrium:

$$P + F = C + 2 \tag{1}$$

where P is the number of phases, C is the number of components in the system, and F is the variance, or number of degrees of freedom of the system. These terms and their significance will be discussed in turn.

Selection of Components: Application of the Phase Rule

To apply the phase rule, the number of components (C) has first to be determined by inspection and analysis of the system. For example, an early stage of reaction in cement making is the decomposition of calcium carbonate ($CaCO_3$). Decomposition proceeds according to the equation:

$$CaCO_3 = CaO + CO_2 \qquad (2)$$

$CaCO_3$ and CaO are solids and CO_2 is gaseous under normal conditions of the reaction. Therefore the simplest choice of components is two: CaO and CO_2. This is because $CaCO_3$ can be regarded as comprised of 1 mol each of CaO and CO_2. Note that it would not be wrong to describe the system in terms of the three components Ca, C, and O, but this choice would complicate the subsequent phase rule analysis because we would have to include substances such as calcium carbide and carbon monoxide, which, in practice, do not appear. Thus the preferred choice of components is the simplest that correctly embraces the compositions of all the reactants and products. This choice of components may not always be intuitively obvious; considerable knowledge of the reactants and their products may be required to justify the selection.

Another example suffices to make this point. When cements are hydrated, we normally assign the iron oxide component to Fe_2O_3. But in cements containing certain slags or fly ash, and in high-alumina cements, iron is partly present in chemically reduced form (Fe^{2+}) and it may be necessary to treat the iron as two components: under these circumstances FeO and Fe_2O_3 are the usual choice. But if iron metal were also known to be present, as occurs when embedded steel reinforcement is included as part of the system, it might be preferable to select iron and oxygen as components: FeO could then be regarded as composed of ($Fe + 0.5O_2$). If iron appears as a hydrated iron oxide phase or phases, H_2O has to be added as an additional component, but, since we are considering cement hydration, water would in any event be selected as a component.

The large number of components present in hydrating cement could potentially give rise to a large number of solid phases. For example, a simplified cement paste may be regarded as consisting of five components: CaO, M_2O_3 (including Al_2O_3 and Fe_2O_3 as one component, which under some circumstances may not be appropriate), SiO_2, SO_3, and H_2O. Ignoring the vapor phase and applying the condensed phase rule ($P + F = C + 1$) gives $P + F = 6$. Thus if $F = 0$, the maximum number of coexisting phases would be six: five solids and an aqueous phase. Note, however, that the application of the phase rule gives the maximum number of phases coexisting at equilibrium — it does not state that we must have the maximum. Fewer than the maximum are not excluded and, in any event, F may be greater than zero; the system need not be invariant. Examination of portland cement pastes discloses the existence of four persistent solids: $Ca(OH)_2$, C-S-H with Ca/Si close to 1.8, AFt (ettringite-like phase), and AFm. These phases appear across a range of temperatures close to 20°C, suggesting that the system must retain at least one degree of freedom. Thus the phase analysis is not inconsistent with the phase rule. Note however that C-S-H is metastable in the absolute thermodynamic sense. The system contains a persistent assemblage consisting of stable and metastable phases. (The special roles of metastable phases (e.g., C-S-H and AFm) will be discussed subsequently.) Of course this analysis is only approximate: experience suggests, for example, that the crystal chemistry of iron and aluminium oxides differ in a hydrated cement paste and that each oxide, Fe_2O_3 and Al_2O_3, may have to be treated as a separate component. But complex systems may require simplification in the first instance; provided the basis of simplification is stated and does not conflict with real life, this is acceptable.

Graphical representations of multicomponent systems remain a problem. Systems of one, two, and even three components can be represented graphically without much need to restrict the amount of information contained. But no amount of ingenuity can devise completely satisfactory schemes to depict systems of four or more components; it is generally necessary to rely on the constituent ternary systems together with sketch maps showing selected relevant features of higher-order systems. However, and as will be shown subsequently, computer-based schemes enable calculation of higher-order systems of four or more components. Phase diagrams still remain an invaluable introduction and guide to phase relationships among the phases, although, increasingly, they will be supplemented by calculations.

If other components are added to the simplified system described in the preceding paragraph, how will the phase rule description be affected? Broadly, other components can be divided into two categories: those that are soluble, and those that are not. Thus sodium and potassium at normal concentrations either dissolve in pore fluid or undergo sorption on cement substances (e.g., C-S-H). No new phases are formed at alkali levels up to at least ~1 M in the pore fluid. On the other hand, magnesium, perhaps introduced in the clinker or in blending agents, is insoluble in the other phases, solid as well as aqueous, and, because it does not substitute significantly for calcium in cement solids, magnesium must appear as an additional solid phase or phases, for example, brucite, $Mg(OH)_2$, hydrotalcite, M_4AH_{10}, and/or M-S-H gel. Thus an operational distinction can be made.

Cements are made from naturally occurring and impure raw materials, and, moreover, are used in service conditions such that they are liable, or potentially liable, to react with a range of chemical species that may be present in mix water or in the service environment. Do these chemical species, whether present in clinker or mix water, or introduced from the environment, count as components? A pragmatic answer is that if a chemical species affects the nature of reaction, it needs to be included. Thus, for example, sea water is a relatively dilute (~0.05 M) solution with respect to its dissolved magnesium content and, at first sight, it might appear acceptable to neglect magnesium as a component in assessing the thermodynamics of the interactions of sea water with low-magnesium cements. But field studies, supported by laboratory experiments, disclose that the magnesium component of sea water reacts strongly with cement paste, forming solid hydrates containing essential magnesium.[4] However, if the magnesium is present in mix water, as occurs if sea water is used to make concrete, the total mass of magnesium per unit volume of cement is low; formation of magnesium-containing solids must consequently be very restricted in amount and can be neglected. But if concrete is immersed in sea water, although the concentration of magnesium in sea water is still low, the sea water furnishes an effectively infinite and inexhaustible source of magnesium with the result that magnesium must be included in a thermodynamic analysis of sea water attack on cement paste.

Furthermore, if we seek to reduce the number of components by including magnesium as a specific salt, for example, as magnesium sulfate (i.e., restricting thermodynamic analysis to equal atom ratios of Mg and SO_4), the approximate nature and artificiality of the assumption must not be forgotten in the course of subsequent analysis. The molarities of Mg and SO_4 in sea water are not equal and can differ greatly among natural brines and groundwaters. Moreover, in the course of laboratory testing of cement monoliths in $MgSO_4$ solution, magnesium is typically removed from solution by reaction faster than sulfate. Although concentrations of the two species may be equal at the outset, as occurs if the solution is formulated using $MgSO_4$, the two concentrations (Mg and SO_4^{2-}) rapidly and spontaneously become unequal as reaction progresses.[5] Thus in the general case it may not even be appropriate to combine the two species Mg and SO_4 as a single component.

While the large number of admissible components and partial openness of the system might at first sight appear to greatly complicate the application of thermodynamic methods to deterioration processes arising as a consequence of mass fluxes into and out of concrete, the system concept also enables us to deconvolute complex, multicomponent systems into subsystems, each of which can be studied in its own right. For example, if we were to analyze the impact of NaCl on hydrated cement paste, we might determine separately the impact of NaCl on the $CaO-Al_2O_3-H_2O-NaCl$ and $CaO-SiO_2-H_2O-NaCl$ systems, the first mainly to elucidate the reactions of aluminates with NaCl and the second to determine the impact of NaCl on the silicate phase(s).[6-8] This done, calculations could be done on the system containing both aluminate and silicate. Indeed, the majority of thermodynamic studies thus far published on cements concern phase relations in subsystems. It should in theory be possible to assemble the subsystems into a continuous framework; examples of the integration process are given below. Because of the complexity and openness of environmentally conditioned reactions, most of the subsequent presentation will concentrate on closed-system behavior.

The variance of the system (F) is best illustrated with reference to a simple example such as that shown in Fig. 2 for $CaCO_3$ decomposition. Both pressure and temperature are important; in this reaction the relevant pressure is that of carbon dioxide. Thus the temperature at which decomposition occurs is related to CO_2 partial pressure and vice versa. At atmospheric P_{CO_2} (~5 × 10⁻³ bar)... At atmospheric P_{CO_2} (~5×10^{-3} bar), the decomposition temperature, normally 885°C for 1 bar CO_2 pressure, is significantly lowered. Likewise, the decomposition of a hydrate such as $Ca(OH)_2$ occurs at a temperature defined by water vapor pressure; alternatively, if temperature is fixed, the water vapor pressure for decomposition is fixed. For more complex substances, such as ettringite, decomposition may not occur under strictly equilibrium conditions; formation and decomposition reactions are not strictly reversible.[9,10] However, in much of cement hydration, the stable phase assemblages are relatively insensitive to pressure with the result that total pressure is not an important variable. Autoclaved cements and cements in deep, hot wells are important exceptions to this generalization, although in these examples both tempera-

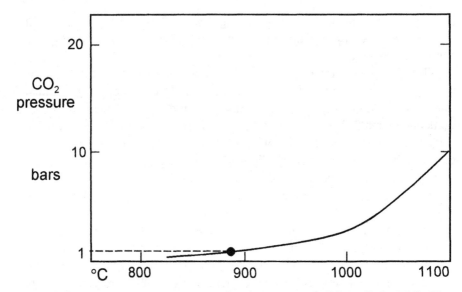

Figure 2. Pressure vs. temperature curve for the reaction $CaCO_3 = CaO + CO_2$. The dashed line, at $P_{CO_2} = 1$ bar, defines the "normal" decomposition temperature given in handbooks, 885°C, and is shown by a dot.

ture and pressure lie beyond normal limits for constructional concretes. If, however, pressure remains constant (i.e., if isobaric conditions obtain), the phase rule can be restated in simplified form:

$$P + F = C + 1 \tag{3}$$

In many practical applications, temperature is also effectively fixed by service conditions, thus providing another source of restraint. Indeed, cement and concrete often experience approximately isothermal service conditions, perhaps in the range 0–40°C. Thus property data obtained on cements often relate to a central value in the range 20–25°C. However, data thus obtained should be applied with caution at other temperatures; examples will be discussed subsequently in which the thermodynamic relationships may change drastically, even within a small range of temperatures, 0–40°C. Thus, while thermodynamic data obtained at one temperature can in principle be transformed to other temperatures, such transformations need to be subject to a reality check.

The phase rule also helps in defining boundary conditions. Consider water as an example. Experience suggests that water behaves as a one-component system because it decomposes to its elements only at extremely high temperature. However we normally need to apply the full form of the phase rule (Eq. 1), because water has an appreciable vapor pressure over the range of temperatures of interest. But because composition remains constant, we can use one axis of a conventional two-dimensional graph to depict temperature and use the other for pressure. Presenting the phase diagram for H_2O in this way (Fig. 3) displays fields of liquid, vapor, and solid. Boundaries between the different phases are readily apparent. An interesting and perhaps unforeseen circumstance is the occurrence of not one but several crystallographically distinct forms of solid water, termed polymorphs. While ordinary ice is the most familiar polymorph of the solid, other polymorphs have stability fields depending on pressure and temperature. While chemically identical, polymorphs differ in physical properties such as density, refractive index, and crystal structure. Thus each polymorph has a distinctive set of thermodynamic properties. If we wish to determine the coexistence of these polymorphs, the maximum number of phases will occur when F = 0, that is, when pressure and temperature are both fixed. Under these conditions and applying the full form of the Gibbs phase rule (Eq. 1), P must be 3. Thus, for example, we can have at most one solid phase coexisting with both vapor and liquid. This is a so-called triple point, which, in this instance, marks the minimum temperature for the stable coexistence of liquid water with both ice and vapor at approximately 0°C. The locus of a triple point in a one-component system is therefore defined by the intersection of three univariant curves, that is, curves for which F = 1. For example, the univariant curve separating the fields of vapor and liquid defines the well-known dependence of the boiling point of water with pressure. In order to specify completely the state of the system with one degree of freedom, as occurs on any of the unvariant curves, this degree of freedom must be fixed by defining either pressure or temperature. Thus points on the liquid-vapor curve define the dependence of the boiling point of water as a function of water vapor pressure. Inspection of the diagram shows two other types of invariant points: one defined by the existence of two solids with liquid, the other by three solids. Although the univariant curves have definite terminations,* nonequilibrium extensions of invariant curves are also possible. For example, considering the water-ice-

---

*The case of the termination of the liquid-vapor curve (not shown in Fig. 3) at a critical point is a special example that is not discussed here.

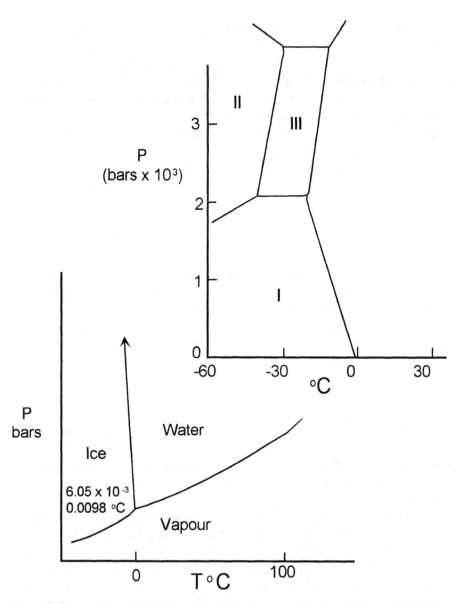

Figure 3. Phase diagram for H_2O as a function of temperature and pressure. The inset diagram, lower left, shows low pressures, including numerical values of the pressure (bars) and temperatures of the triple point marking the coexistence of ice, water, and vapor. Higher-pressure phase transitions in ice are shown in the upper right corner. Note how the freezing point is initially depressed by rising pressure but increases at pressures above ~2000 bar.

vapor point, water can readily be supercooled to below its normal freezing point at 0°. However, this situation will not persist indefinitely: if bulk water is held in its supercooled state a little below its stable freezing point, ice will nucleate and freezing will eventually occur. The product of freezing in these conditions is still ordinary ice, although at even slight (1–2°C) undercooling, the growing crystals tend to develop unusual morphologies so that experience of ice morphology, or shape, can often be used to infer the conditions that obtained at the instant of freezing. However, if freezing occurs at very large undercoolings (~150–200°C), an amorphous glassy phase (not shown in Fig. 3) can also be obtained.

Cement systems are intrinsically more complex than single-phase systems (e.g., water), on account of the increased number of components and corresponding increase in the number of phases; as noted, the complexity of a system tends to increase factorially with the number of components. But the same principles operate as were described for the water system; boundaries exist that, when crossed, affect the nature of coexisting phases. In the water system, all phases have the same composition; mass transport is not required in the course of phase transformations.

In more complex and heterogeneous systems, mass transport may be required to effect change. Under these conditions, kinetics of diffusion and transport, nucleation, crystallization, and dissolution may have to be coupled with equilibrium to achieve a fuller understanding of the system. But phase boundaries exist in all cement systems, simple as well as chemically complex. They are important in the analysis of cement durability. For example, the thaumasite form of sulfate attack (TSA), which will be discussed below, is speculated in the literature to involve a transition from ettringite to thaumasite as the stable sulfate phase. Although the two minerals ettringite and thaumasite are isostructural, or nearly so, they are known to be incompletely miscible. Therefore we can predict that each phase has special chemical and thermal requirements for its formation and persistence and that small changes in the chemical environment may markedly alter the nature of coexisting solid(s): in particular, minimum aqueous concentrations of the constituent species are required to stabilize particular solids or combinations of solids.

Enthalpy Concepts

As noted earlier, the hydration of cement is exothermic: heat is liberated in the course of hydration. This heat is termed "enthalpy." Enthalpy changes are most accurately measured for cements and cement substances indirectly, from heats of dissolution or, more commonly but somewhat less accurately, by isothermal calorimetry. Nonisothermal methods require correction for the heat capacities of all the phases participating in the reaction, solid as well as aqueous, to determine true heats of reaction. In the course of cement hydration, the amounts of each phase change rapidly, and it is difficult to assign heat capacities and enthalpies, numerical values of which are temperature dependent. It is therefore difficult to deconvolute data derived from adiabatic calorimetry, hence the preference for isothermal methods.

Figure 1 shows a representative plot of heat evolution obtained by isothermal calorimetry from a modern portland cement during early hydration; as noted, the rate of heat release follows a complex pattern. A large initial burst of heat evolution, often not well resolved by commercial instrumentation, is followed by a diminished but nearly constant release lasting several hours.

The initial burst of heat evolution, occurring in the first few seconds or minutes of reaction, is often termed a "heat of wetting." This initial heat evolution is often explained by starting with the observation that the surfaces of cement minerals have strong unsatisfied surface charges; such surfaces attract and orient the dipoles of water molecules. This process is normally exothermic, but the attribution of the exotherm to heat of wetting provides only a partial explanation of a complex set of processes, exothermic as well as endothermic. During the first few minutes of hydration, much soluble matter is dissolved from the cement, including alkali sulfates, calcium from cement minerals, and so on. As a consequence of this dissolution, the aqueous solution becomes strongly alkaline, the pH rising from near-neutral to 12 or more. The resulting increase in aqueous hydroxyl ion concentration, together with the presence of other dissolved species, results in physical changes to the "water": for example, its ionic strength and density change. Chemical changes also occur in the extent of ionization of water molecules and of the dissolved species, resulting in corresponding changes to the strength of dipole-molecule and dipole-dipole interactions within the solvent. These result in rapid enthalpy changes to the "water" during the initial stages of hydrating and cumulate with mineral surface–water interactions to give the so-called heat of wetting. Thus "wetting" refers to only one of many processes occurring as cement is mixed with water. To the author's knowledge no proper modern study has been made of the contributions of the separate processes to the overall energy balances occurring during the first few minutes of reaction.

Table I. Enthalpy of hydration of cement components

Compound	Reaction conditions	ΔH (kJ/kg)
C_3S	$+H_2O$, yields C-S-H and CH	-517
β-C_2S	$+H_2O$, yields C-S-H and small amounts of CH	-262
C_3A	$+H_2O$ and gypsum: product is $C_4A\bar{S}H_{12}$	-1144
C_3A	$+H_2O$ and gypsum: product is ettringite	-1672
C_4AF	$+H_2O$ and CH: product is mainly hydrogarnet	-419

Enthalpy changes are additive, so a firm conclusion is that, overall, the sum of the enthalpy changes occurring in the course of the first few minutes of reaction is strongly exothermic; that is, heat is removed from the system to its surroundings. By convention this change in system enthalpy is negative. Subsequent to this initial exothermic event, a period of relatively slow heat evolution occurs, sometimes referred to as the "dormant" period. Electron microscopy suggests that this period of slow heat evolution is associated with formation of a semi-protective solid film on clinker grains, which is believed to slow mass transport, especially of water, and thus hinder reaction. Eventually this film is disrupted and reaction again becomes rapid as liquid water contacts anhydrous clinker. Thus the rate of observed heat output is clearly subject to kinetic controls: one speculative explanation of the mechanistic origin of these controls is discussed under osmotic pressures.

Further enthalpy changes occur, albeit at diminishing rates, as hydration approaches completion. If we consider the total heat evolved, it will be the sum of heats of the individual reactions of the constituent phases with water (and, where appropriate, with each other), taking into account the mole fraction of each clinker phase. This is a statement of Hess's law applied to cement hydration. Hess's law (Eq. 4) states that the overall enthalpy change is the algebraic sum of the enthalpy changes of the constituent processes, taking into account the mole fraction of each component. Thus the overall progress of hydration can also be correlated with the progress of hydration from knowledge of the heats of hydration of the constituent clinker phases.

The constituent minerals of cement clinkers have been synthesized and their individual total heats of hydration measured with respect to formation of specific hydrates. Table I gives consensus values probably accurate to $\pm 5\%$. Other measures of heat releases, such as heat release as a function of time (Table II), are useful but less fundamental because the data in Table II

Table II. Heat evolution as a function of time*

Phase	\multicolumn{6}{c}{Hydration duration (days)}					
	3	7	28	90	180	365
C_3A	7.1	7.5	8.0	8.4	8.8	9.2
C_3S	3.3	3.8	4.2	4.6	5.0	5.3
C_2S	0.4	0.8	1.3	1.7	2.1	2.5
Ferrite	0.8	1.3	1.7	2.1	2.5	3.0

*J/g per 1% of the four clinker minerals in a portland cement. Temperature = 23°C and w/c = 0.45. The specific surface (Blaine) is assumed to be 3100 ± 200 cm²/g.

are functions of water/cement (w/c) ratio and surface area, as well as of particle size distribution and therefore this type of data presentation conveys information mainly on the interactions of kinetics with equilibrium, although deconvolution of the separate factors is not readily achieved.

The relevant hydration stoichiometry of silicates is essentially fixed; C_3S in cements behaves the same as "laboratory" C_3S, although for the aluminates and aluminoferrite phases reaction stoichiometry depends critically on the total amount of sulfate and its availability to participate in hydration reactions. For example, at low sulfate availability, a hydroxy AFm or a mixture of hydroxy and sulfate AFm phase are the preferred products of hydration of aluminates, whereas at high sulfate availability ettringite (AFt), forms. In practice, and as reaction proceeds, most portland cements contain insufficient sulfate totally to combine all M^{3+} ions (Al, Fe) as ettringite, with the result that after initial formation of AFt, AFm becomes favored as both supply and availability of sulfate reduce. For the aluminoferrite phase, which in any event tends to hydrate slowly and mainly in a sulfate-deficient regime, mixtures of AFm and hydrogarnet form as the preferred products of hydration. Thus it is important to use the appropriate reactions to calculate heat release attending hydration. Fortunately, not only does ferrite hydrate slowly but its heat of hydration is also low in all its principal reactions, so the accuracy of enthalpy release calculations depends mainly on aluminate content and less on ferrite content and its hydration behavior. For calculations appropriate to the first few days of hydration, such that sulfate reactant is still present, reaction stoichiometrics yielding AFt (ettringite) are usually used, reflecting the experimental observation that ettringite is the main aluminate hydration product encountered at short ages.

Enthalpy changes in a hydrating cement clearly involve a series of reactions; as noted, the enthalpies of the separate reactions are additive. Thus, for example, if Q is the heat liberated at full hydration and assuming the cement contains negligible free lime or magnesia, Hess's law can be expressed as:

$$Q_{total} = q_1(C_3S) + q_2(C_2S) + q_3(C_3A) + q_4(C_4AF) \qquad (4)$$

where Q is total heat evolved and q_1 through q_4 are the separate contributions from each clinker constituent, the mole fraction of which is either known, measured, or calculated (e.g., using a Bogue-type calculation).

The q values selected for C_3A and C_4AF will, as noted, depend on sulfate content and availability over the duration of the time period being calculated. At ~20°C and long time periods, such that hydration is complete, ettringite is the preferred product at high sulfate contents, C_4AH_x* at zero sulfate, and AFm, or mixtures of AFm with either C_4AH_x or ettringite, at intermediate sulfate contents and/or availability. The sulfate balance is important as, in general, C_3A releases more heat per gram than either C_3S or C_2S and the nature of its reaction is, as noted, sensitive to sulfate content. Thus enthalpy data for the constituent reactions can be used to predict with reasonable (±5–10%) accuracy the overall energy release attending hydration and, coupled with kinetic data, the heat released over selected time intervals.

The reaction enthalpies lie at the heart of designing low heat of hydration cements. Inspection of Table I discloses that the most effective way of lowering heats of hydration is to control clinker mineral content; lower C_3A and C_3S contents lead to lower specific releases. These data underlie all specifications for low-heat portland cements. Coarse clinker grinding does not in itself affect total heat release, but, other factors being equal, it spreads the heat release over longer time periods with the result that a concrete made with a coarser-grained cement will experience a lower temperature rise than its fine-grained equivalent.

The actual temperature rise experienced within a mass of concrete depends on the heat of hydration per unit time as well as a host of other factors, including concrete composition, geometry, and the physical size of pours as well as the thermal conductivity of the mass, the heat capacity (including that of the aggregate), and the start temperature at the onset of hydration. Since this type of calculation depends in part on extra-thermodynamic quantities it

---

*Depending on the activity of water (see below), x is typically 13 or 19.

is not pursued here, but, ultimately, engineering calculations must utilize Hess's law, coupled with basic thermochemical data for the individual hydration reactions and the heat capacities of the components, including aggregates, and, of course, make allowance for heat loss by conduction.

Kinetic Considerations and Enthalpy Changes

Studies of the hydration kinetics of the individual constituent mineral phases of cement disclose that, other factors being equal, they react at different rates with water. It might therefore be supposed that, since alite and tricalcium aluminate have high enthalpies of reaction and, moreover, are known to hydrate faster than either belite or ferrite, complex kinetic factors would have to be introduced into Eq. 4 to allow for the impact of differential rates of hydration on the cumulative pattern and the time dependence of heat evolution. In practice, it is not usually necessary to introduce corrections for differences in reactivity of the constituent phases except possibly during the first few hours or days of hydration. The microstructure of hydrating clinker grains made to normal w/c ratios reveals that several hydration mechanisms exist, but often a layer of relatively dense hydration product accumulates at the grain surfaces. This accumulation of hydrate products restricts access of water to fresh, unreacted clinker and thereby becomes the rate-limiting factor for hydration. Thus the clinker gains in a paste made to "normal" w/c ratios follows a shrinking sphere model (treating clinker grains as approximately spherical particles) in which the unhydrated core continues to erode almost uniformly, albeit more slowly, as hydration proceeds and the semi-protective hydrate layer thickens. The onset of these conditions, appropriate to application of a shrinking sphere model, coincides approximately with final set. In the post-set condition, little differential reaction of the separate minerals occurs and with limited water access to the remaining solid anhydrous minerals, Ca_3SiO_5, Ca_2SiO_4, and $Ca_3Al_2O_6$ tend to hydrate at approximately the same rate.* The principal exception is ferrite, crystals of which frequently remain embedded in otherwise well-hydrated cement paste as unreacted or partially reacted relicts. However, the enthalpy change associated with ferrite hydration is low and, as noted, its hydration rate does not therefore much influence numerical calculations of the overall heat release.

---

*Clinkers containing belite clusters may be an exception.

Entropy Concepts

Energy does not necessarily appear only in the form of sensible heat. For example, at all temperatures above absolute zero the constituent atoms of a solid, liquid, or gas have a thermal amplitude of vibration that absorbs energy, the amount of this energy increasing with temperature. This energy is reflected in the observed heat capacity, and because the energy-absorbing modes become saturated explains why the heat capacity tends to approach a constant limiting value. Changes in state — for example, polymorphic transformations in solids, melting, boiling, freezing, and condensation — all result in additional entropy changes: energy is absorbed or liberated at constant temperature. Thus when water converts to ice at 0°C, or vice versa, heat is absorbed or liberated, depending on the reaction direction, although the temperature remains constant until the energy necessary to complete the transformation has been exchanged. This energy is termed "entropy." Thus, in considering the overall changes in energy in a system, both enthalpy and entropy need to be taken into account. This is true even under isothermal conditions. For example, at 0°C the entropy of a water molecule in ice or in a crystalline hydrate differs from that of a water molecule in bulk water. These differences affect the colligative properties of the bulk material. For a one-component system, the distinction between enthalpy and entropy is often obvious, but in more complex systems the distinction may be less obvious.

Historically, it has been tempting to take a negative enthalpy as the criterion for a spontaneous reaction. After all, we have seen that hydration of cement occurs spontaneously and that it has a negative enthalpy; that is, it liberates heat in the course of hydration. Additional examples can be offered. If we dissolve solid sodium hydroxide in water — clearly a spontaneous process — much heat is evolved (negative enthalpy, according to convention). So far, so good. However, we can also find many salts (e.g,. nitrites) that are spontaneously soluble in water yet absorb heat in the course of dissolution, and as a result of which, adiabatic solution temperatures decrease. The adiabatic temperature change attending dissolution can thus be either positive or negative; enthalpy changes alone are insufficient to distinguish. We note that when ΔH, the enthalpy change, is numerically larger than the entropy change, ΔS, its sign and numerical value tend to dominate the overall observed behavior; indeed, this condition characterizes the majority of inorganic reactions. But the examples given above also lead to the conclusion that the spontaneous nature of reactions must find a more fundamental expression. This lies in the concept and definition of "free energy" — not in this author's view an appropriate term, but one that has historical precedent.

Gibbs Free Energy

Since we are mainly concerned about changes in energy occurring at constant pressure or over small ranges of pressure, we use the Gibbs definition of free energy (G). Again, we are more concerned about energy changes (Δ) than absolute values, in which case the Gibbs free energy of a reaction may be defined as:

$$\Delta G = \Delta H - T\Delta S \qquad\qquad (5)$$

where T is temperature in Kelvin and ΔH and ΔS are, respectively, changes in the enthalpy and entropy. The true criterion for a spontaneous reaction is that the numerical value of ΔG should be negative; it is the measure of the driving force for reaction. Calculations involving changes in ΔG are relatively easy; like enthalpies, free energies are additive, taking into account reaction stoichiometries and arithmetic signs. Large overall negative values of ΔG mean that a reaction goes essentially to completion, but small values imply that significant concentrations of products and reactants may remain at equilibrium.

An example illustrating the free energy changes attending hydration of cement at ordinary temperatures (~20°C) is shown in Fig. 4; the left side depicts energy changes at ~1500°C. At 1500°C formation of Ca_3SiO_5 is energetically favorable, although at ~25°C it is unstable with respect to Ca_2SiO_4 and CaO. Ca_3SiO_5 is used as a model substance and the overall ΔG changes are qualitative. A mixture of Ca_3SiO_5 (solid) and H_2O (liquid) is again taken as an arbitrary zero and a large negative ΔG change occurs upon hydration of alite to yield a mixture of $Ca(OH)_2$ (solid) and C-S-H (solid). If we were to quantify the calculation, it would be necessary in the first instance to write a balanced reaction. This is not a problem in respect of $Ca(OH)_2$, H_2O, or Ca_3SiO_5, all of which have definite stoichiometries, but it is a problem with respect to C-S-H: What is its formula? C-S-H, as is well known, varies in Ca/Si ratio. Moreover it is not rigorously possible to distinguish bound water from capillary, or weakly bonded, water. The Ca/Si ratio of C-S-H coexisting with $Ca(OH)_2$ is high, at or near its maximum possible value. Also, it is not necessary to have an absolute distinction between the different types of water present in C-S-H in order to define the free energy of C-S-H, provided the basis of calculation is clearly stated.

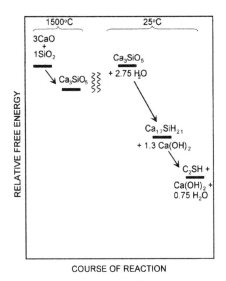

Figure 4. Schematic diagram of energy changes in the system $CaO-SiO_2-H_2O$ illustrative of formation of Ca_3SiO_5, and its subsequent hydration forming C-S-H and, finally, crystallization to the stable state.

Thus, using the data of Young and Hansen,[11] the reaction stoichiometry can be balanced:

$$C_3S + 2.4\,H_2O = Ca_{1.7}SiH_{2.1} + 1.3\,Ca(OH)_2 \quad (6)$$

Of the two solid products of reaction in Eq. 6, one — $Ca(OH)_2$ — is crystalline while the other — C-S-H — is amorphous, or nearly so. If C-S-H were to crystallize — and it is possible to envisage different products of crystallization, depending on temperature and C-S-H composition — a further overall decrease in free energy would occur, as shown in Fig. 4. As is well known from the persistence of semi-amorphous C-S-H, both in its natural occurrence as a mineral and in cement paste, crystallization is kinetically slow at ambient or slightly elevated (<50°C) temperatures, and we often wish to base calculations on the metastable but persistent phase assemblage of C-S-H and $Ca(OH)_2$. Thus, quantification of the numerical values of ΔG depends on a number of other factors, including definition of reaction stoichiometry, but the pictorial representation shown in Fig. 4 forms a useful adjunct to more complex calculations. The very large and favorable free energy change attending hydration suggests, for example, that the reaction of Ca_3SiO_5 and water will tend to go to completion at ~20°C; that is, all Ca_3SiO_5 will tend to be consumed, given a sufficiency of water. We return to quantification of this reaction subsequently.

However for other reactions the driving force, ΔG, may be small. For example, in the $CaO-Al_2O_3-SO_3-H_2O$ system at 1 bar pressure and ~25°C, ettringite is stable with respect to sulfate AFm, but the additional free energy attending formation of ettringite from AFm is slight: indeed, sulfate AFm becomes thermodynamically stable at or above ~40°C. The energetic balance of this type of reaction is expressed by the law of mass action. For example if four solids, designated A, B, C, and D, are related by the reaction:

$$A + B = C + D \tag{7}$$

the law of mass action states that at a fixed temperature and pressure, the velocity of reaction is proportional to the arithmetic product of the active masses of the reacting substances. If k_1 and k_2 are the velocities of the forward and reverse reactions, respectively, and if brackets are used to indicate the reactive masses,

$$v_{(forward)} = k_1 [A] [B] \tag{8}$$

and

$$v_{(reverse)} = k_2 [C] [D] \tag{9}$$

Combining these equations gives

$$k_1 / k_2 = \frac{[C][D]}{[A][B]} = K \tag{10}$$

where square brackets indicate concentration. Strictly, aqueous concentrations should be in molal units, although at high dilutions, such that differences between molar and molal scales become small, molar units can often be used without serious error. For reactions where the stoichiometric coefficients are not unity, e.g., for reactions of the type:

$$aA + bB = cC + dD \tag{11}$$

The appropriate expression must introduce the stoichiometric coefficients a, b, c, and d and becomes:

$$K = \frac{[C]^c [D]^d}{[A]^a [B]^b} \tag{12}$$

These are examples of concentration equilibrium constants (K); a brief digression is appropriate to note that other types of K values can be defined, for example, in gas equilibria, where fugacities or partial pressures may be used as effective measures of concentration. The K values are usually designated according to the concentration units; for example K defined in terms of pressure is designated by K_p.

K values thus quantify the tendency toward completeness of a given reaction and ΔG, the driving force, may also be cited as a measure of com-

Table III. Thermodynamic stability of some AFm-type phases at 20°C

Anion substituent and name	Stable?
OH (hydroxy AFm)	No
SO_4 (sulfate AFm)	No*
Cl (Friedel's salt)	Yes
CO_3^{2-} (hemicarboaluminate)	Yes
CO_3^{2-} (monocarboaluminate)	Yes

*Stable above ~40°C.

pleteness of reaction. ΔG, the free energy, is related to equilibrium constants by the equation:

$$\Delta G = -RT \ln K \qquad (13)$$

where T is temperature in Kelvin, R is the gas constant, and K the equilibrium constant. In subsequent applications we make extensive use of solubility constants and their relation to free energy.

As a rule of thumb, a reaction occurring at ~300 K and having ΔG values less than ~5 kJ/mol will tend to give a mixture of products and reactants; that is, the reaction will not go to completion. The coexistence in hydrated cement paste of AFm phases with ettringite, hydrogarnet, and so on was previously cited as an example of incomplete reaction; although AFm is believed to be metastable with respect to assemblages including AFt, it is persistent. This arises partly as a consequence of the low driving force for reaction (only a few kJ/mol at 25°C), but two other factors are relevant. First, the exact numerical value of ΔG for formation is influenced by chemical constitution; for example, AFm has a complex chemistry and we know that some chemical substitutions stabilize AFm in the thermodynamic sense. For example, substitution of carbonate for hydroxyl in AFm, resulting in formation of calcium hemi- or monocarboaluminate, will stabilize an AFm-type phase across the range of temperatures relevant to cement hydration. Table III shows how the thermodynamic stability of AFm phases at ~20°C depends critically on anion content.

Another reason for the persistence of AFm is that the delicate balance of free energies is affected by solid solution; anion substitution in AFm is given as an example. Table III gives data only for end-member compositions, but in real life, AFm phase chemistry is characterized by extensive solid solution between chemically different anion species and, in cases such that solid solution is incomplete, formation of ordered compounds may occur (e.g., monocarboaluminate) containing both OH and CO_3 in structurally ordered positions. In the first case, where anions are disordered or have only limited site preferences for occupancy, random distribution of several types of anions among the available sites often gives rise to an additional small stabilization energy owing to a favorable entropy of mixing contribution.

Changing the focus slightly, solubility concepts can be more clearly illustrated by the use of $Ca(OH)_2$ as an example. Its solubility reaction implies an excess of water and may be written as

$$Ca(OH)_2 = Ca^{2+} + 2\,OH^- \tag{14}$$

and

$$K = \frac{[Ca^{2+}][OH^-]^2}{[Ca(OH)_2]} \tag{15}$$

This simplifies to

$$K_{sp} = [Ca^{2+}]\,[OH^-]^2 \tag{16}$$

This special type of K in Eq. 16 is termed a "solubility product" (K_{sp}). Note that the solid remains constant throughout, so it need not appear in the expression defining K_{sp}, and that the equation defining K_{sp} has to be stated in all but the simplest cases, as its definition may not be intuitive as in the above. Indeed, in common with other salts with divalent cations, the ionization of $Ca(OH)_2$ actually proceeds in steps, and we might wish to distinguish the two, in which case we need to write two reactions, each of which would have a separate K, thus:

$$Ca(OH)_2 = Ca(OH)^+ + OH^- \tag{17}$$

$$Ca(OH)^+ = Ca^{2+} + OH^- \tag{18}$$

The sum of Eqs. 17 and 18 is, of course, equal to Eq. 14.

Since the ionization of $Ca(OH)_2$ is essentially complete under most conditions, including especially the self-generated pH of its saturated solution, the low aqueous concentration of $Ca(OH)^+$ can be ignored. Thus we often define and use a single K_{sp}, combining the two ionization steps to obtain a single overall reaction, although accepting that the process of dissolution occurs in stepwise manner.

In other cases it may be important to define the reaction in order to express speciation, the form in which a particular entity appears in solution. Carbon dioxide is a case in point. As the pH of a solution increases, dissolved CO_2, present in acid or neutral solutions as CO_2 or H_2CO_3, converts to HCO_3^- and, at high pH, to CO_3^{2-}. Thus mutivalent anions, like cations, may ionize in stages. For CO_2 only the last two speciations need be used to describe reactions in alkaline media, and of these two species, only one, CO_3^{2-}, is important at pH values around 12.[12] A further example involves Al, which tends to appear in acid solutions as Al^{3+}, or as a hydrated Al species, but appears in alkaline solutions typically as the aluminate ion $Al(OH)_4^-$. The form of writing species affects the balance of the equation and hence the numerical value of K. Thus if we are given a K_{sp} for ettringite, it is important to define the form of the dissolution equation. It might, for example, be written

$$K_{sp} = [Ca^{2+}]^6 \, [Al(OH)_4^-]^2 \, [OH^-]^4 \, [SO_4]^{3-} \tag{19}$$

But the dissolution equation could also be balanced showing aluminium as Al^{3+}, or $Al(H_2O)_6^{3+}$, and so on. The choice of species would affect the definition of K_{sp} as well as its numerical value. Note, however, that because water is the solvent, it is not necessary to include water in the defining reaction.

The reader may well ask: How am I to decide which speciation to use? The answer is not straightforward. Computer iterations will generally utilize built in speciation data, thus apparently avoiding the problem. However, in order to have written the computer program in the first place, or for the user to understand the results and limitations of calculations, it may be necessary to acquire a broad general knowledge of speciation. A general textbook on aqueous chemistry or geochemistry will supply sufficient information for most applications. It is important to realize that there is often no right or wrong formulation (although some choices may be inappropriate), provided the defining reaction is clearly stated.

Many calculations using the K_{sp} concept can be undertaken. For example, we may wish to determine a species solubility from a K_{sp} or vice versa. To keep the calculation simple, portlandite, $Ca(OH)_2$, is used as an example. Given its K_{sp}, 7.88×10^{-6} at 25°C, what is the calcium concentration at saturation? The equation can be solved as follows. Since each calcium in solution requires two hydroxyl ions for change balance, Eq. 14 simplifies to:

$$K_{sp} = 4 \, [Ca^{2+}]^3 \tag{20}$$

Solving,

$$7.88 \times 10^{-6} = 4 \, [Ca^{2+}]^3 \tag{21}$$

and Ca = 2×10^{-2}, or 20 mM (millimolar). If we want to calculate the pH of a saturated solution, it is apparent by inspection that the hydroxyl ion concentration will be twice the calcium concentration (2×20, or 40 mM), assuming complete ionization. From the definition of pH as equal to $-\log$ [H$^+$] and recalling that the ionization product of water, K_{water}, is 10×10^{-14} = [H]$^+$ [OH]$^-$ at 25°C, the pH is readily calculated to be about 12.4.

The concept can be extended to more complex systems. Supposing sodium hydroxide is also present in the aqueous phase. What will the impact of NaOH on portlandite solubility be? Assuming that the sodium concentration is 0.1 M and that other competing anions are largely absent or precipitated (as is normally the case), and noting that sodium is not absorbed on Ca(OH)$_2$, it follows that the charge on sodium must be balanced in solution by hydroxyl. Thus we initially assume [OH$^-$] will also be 0.1 M, justifying this approximation by noting that from the preceding calculation, the OH concentration of a saturated portlandite solution was only 40 mM, that is, much less than 0.1 M. Taking the OH concentration as 0.1 M,

$$K_{sp} = 7.88 \times 10^{-6} = Ca \, [0.1]^2 \tag{22}$$

where square brackets indicate concentration. Solving, the calcium concentration is calculated to be reduced to 0.08 mM in 0.1 M sodium hydroxide. Thus calcium solubility is greatly reduced relative to its solubility in initially pure water. Moreover, any error in the calculation arising by neglecting OH produced from portlandite dissolution must be slight, since [OH] will remain close to 0.1 M. While the calculation could, if desired, be refined by an iterative process to eliminate error introduced by the simplification used to solve Eq. 22, the answer is sufficiently accurate for most purposes. It illustrates the well-known tendency of soluble alkali hydroxide greatly to diminish the solubility of portlandite in cement pore fluid. Thus the calcium content of the pore fluid of commercial cements is much influenced by dissolved sodium and potassium and is in general much less than from a saturated calcium hydroxide solution. This explains why, in commercial cements containing sodium and potassium, the calcium concentration of pore fluid is typically only 1–2 mM. The example also illustrates that shortcuts may sometimes be useful to achieve approximate numerical solutions. Thus, while the underlying thermodynamics may be exact, solutions to thermodynamic problems frequently employ approximations.

Note also that the calculation is specifically associated with sodium as sodium hydroxide; the calculation would not differ if potassium hydroxide were substituted. However if sodium were initially present as sodium chloride, rather than sodium hydroxide, studies of the system $Ca(OH)_2$-$CaCl_2$-H_2O would be required to determine solubilities. These disclose that at less than 5 M NaCl, approximately, the positive change on Na continues to be balanced almost entirely by Cl^-; chloride is not significantly removed by reaction with $Ca(OH)_2$.[8] While we could probably deduce this from practical experience, detailed impacts of soluble salts on $Ca(OH)_2$ solubility require confirmation by experiment or calculation, or both. The situation with other salts (e.g., Na_2SO_4) is not intuitive and experiments and/or calculations in the system $Ca(OH)_2$-Na_2SO_4-$CaSO_4$-H_2O must be undertaken in order to determine chemical and phase balances and pH, which are determined by the concentrations of soluble or potentially soluble anions, mainly OH and SO_4^{2-}.

The K_{sp} concept is not applicable in simple form to substances that dissolve incongruently, that is, have a different ratio of ions in solution than in the solid. For example, C-S-H having the ratio commonly encountered in portland cement, Ca/Si ~1.8, gives a saturated aqueous solution having Ca/Si > 100. It is an important practical example of an incongruently soluble solid.

An important consequence of incongruent dissolution is that, as dissolution occurs, the bulk composition of the remaining solid changes. In the case of C-S-H, the remaining solid becomes progressively enriched in silica, the least soluble component; as its Ca/Si ratio changes, the K_{sp} of the remaining solid also changes.[13,14] This contrasts with the behavior of congruently soluble substances: When $Ca(OH)_2$ dissolves, both the composition and K_{sp} of the residual solid remain constant, although of course the amount of solid diminishes. We discuss subsequently in more detail the treatment of incongruently dissolving substances.

Problems in applying K_{sp} concepts also arise when weakly interacting salts are included in calculations. For example, we might not expect the presence of dissolved sodium chloride to affect portlandite solubility because NaCl solutions are neutral (pH 7) and because neither Na nor Cl appear in the equation defining K_{sp} for $Ca(OH)_2$. Experimental data showing the impact of NaCl on the solubility of $Ca(OH)_2$ at various temperatures

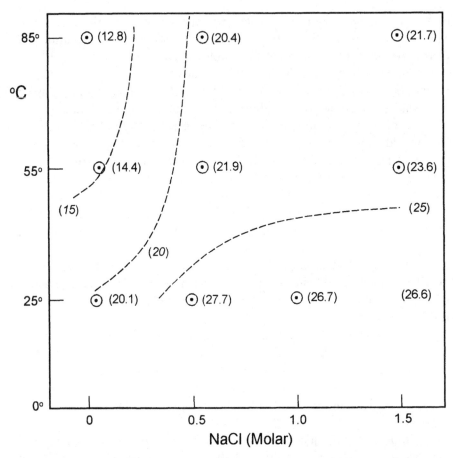

Figure 5. Solubility of Ca(OH)₂ in water and in NaCl solutions up to 1.5 M in the temperature range 25–85°C. Contours have been interpolated at calcium solubilities of 15, 20, and 25 mM calcium; numerical values of contours are in parentheses.

are presented in Fig. 5, demonstrating that this assumption is not strictly true: Significant interactions affecting Ca(OH)₂ solubility occur. Measurements show that the pH of solutions saturated with respect to Ca(OH)₂ and containing 0–1.5 M NaCl remain nearly constant at 25°C, but that Ca solubilities are affected. (Direct comparison at temperatures other than 25°C is difficult because the numerical value of K_w, the ion product of water, is itself temperature dependent). Nevertheless, the same qualitative trends are observed for Ca(OH)₂ solubility in NaCl solutions as in water. The solubility of Ca(OH)₂ in initially pure water decreases with rising temperatures over the range depicted in Fig. 5; this trend probably continues up to ~180°C. The impact of NaCl on Ca(OH)$_2$ solubility is greatest at low NaCl concentrations, <0.5M; at NaCl concentrations in the range 0.5 to ~1.5 M NaCl, higher salt concentrations have progressively less influence on portlandite solubility.

Thus, in the example above, we conclude that ideal behavior is not observed and experimental data may be required to reveal the existence of nonideal interactions that affect solubility, and that if we are unable to deconvolute all the factors and quantify which might be responsible for the impact of NaCl on Ca(OH)$_2$ solubility, experience suggests that the best way forward is to introduce an "activity" factor. Broad experience of aqueous solutions suggests that the nonideal behavior is normally encountered in concentrated aqueous solution. No exact definition of "concentrated" can be given, but, as an approximation, deviations from ideality can be expected at species concentrations greater than 10^{-3} molar. The activity factor conventional symbol γ, is a dimensionless correction factor used to quantify departures from ideality at high solute concentrations. Considerable progress has been made in elucidating species activity–concentration relationships, as will be described subsequently. We could also continue to use K_{sp} concepts in other ways; for example, by introducing a "conditional" K_{sp}, that is, a numerical K_{sp} value appropriate to the solution composition, temperature, and pressure. This might be useful in the special case where cements are exposed to constant environments, as occurs in sea water, but the more general approach of activity coefficients is preferable. Accordingly, much effort has been expended on developing a theoretical basis for activity coefficients.

To develop the activity concept in general form, we need first to introduce the concept of ionic strength (I) in a particular medium (m). With respect to a particular medium of concentration M, the ionic strength (I_m) is given by:

$$I_m = 0.5 \, \Sigma \, M_i Z_i^2 \qquad (23)$$

where M_i and Z_i stand for species concentration and the formal charge on the species (i). The calculation of differences between real and apparent concentrations is thus bridged by introducing an "activity" term (γ). We note that activity coefficients are concentration dependent and, moreover, are not the same for different electrolytes.

However, for a particular valence type, plots of $\log_{10}\gamma$ against I_m tend to become identical at low ionic strengths, as, for example, for all monovalent cations. This is the so-called "limiting behavior." Any successful theory, either theoretical or empirical, has to utilize and explain these observations; both have been pursued vigorously. Perhaps the most successful semi-fundamental approach is the Davies equation, written as follows for a single species:

$$\log 10\gamma = -0.510Z_i^2 \left\{ \frac{\sqrt{I_m}}{1+\sqrt{I_m}} - 0.3I_m \right\} \tag{24}$$

where Z is the formal charge on an ion i at a molal concentration m. The molal concentration scale is preferable to the more familiar molar scale because it enables the concentrations of both solute and solvent to be uniquely defined. The Davies equation takes the charge of ions into account, but not their specific chemical characteristics.

Another, more fundamentally based model for the calculation of solution activity coefficients utilises the Pitzer equations. A derivation is given in Ref. 15. To solve the Pitzer equations, numerical values of a series of coefficients are required; these are available for most neutral species and ions having mono- and divalent charges. The Pitzer approach is often preferred for four reasons: it can cope with mixed electrolyte solutions (e.g., sea water or chemically complex groundwaters), it is computer-friendly, it takes account of the specific chemical characteristics of the species (for example, Mg can be treated differently than Zn), and numerical values of the coefficients for most ions soluble in alkaline solution are known. To solve computer-based calculations, the computer uses input data and accesses a database of numerical Pitzer coefficients, selecting appropriate values of activity coefficients that are introduced into subsequent calculations.

The field of concentration-activity-species relationships remains active, but it is apparent that, owing to the large number of terms needed to describe ion-ion and ion-solvent interactions, computer-based solutions are realistically required for all but the simplest calculations. It is not practicable to introduce a numerical threshold to distinguish dilute and concentrated solutionsm but for most cement-related applications, ideal solution calculations are appropriate only for preliminary calculations.

---

*For simplicity, supersaturation, while permissible, is ignored here.

Partition of Components between Phases

Problems frequently arise in the thermodynamic treatments of cement hydration concerning the distribution of components between phases. In simple cases (for example, of a salt dissolved in a solvent such as water) any solution concentration is in principle obtainable up to the saturation limit.* The saturation limit can of course range over many orders of magnitude; if the solubility is high, as occurs with NaCl, the properties of the aqueous phase (density, freezing point, electrical conductivity, etc.) will be appreciably altered as the concentration of the dissolved species increases. If the solubility is low, as occurs with ettringite, the maximum effect of salt dissolution on aqueous solution properties is diminished. If the dissolving solid is present in excess, the excess remains fixed in composition and its properties are unaffected by the presence of a saturated solution, thus simplifying thermodynamic analysis.

In the case of mixed solutions of calcium sulfate (low solubility) and NaCl (high solubility), complications may arise at high NaCl concentrations, which, by lowering the activity of water, favor formation of hemihydrate ($CaSO_4 \cdot 0.5H_2O$) or anhydrite ($CaSO_4$) rather than gypsum. These minor complications can, however, be treated by the methods thus far described; no new considerations are required. But in heterogeneous systems of several solid phases, or of solids coexisting with an aqueous phase, the solid phases need not have fixed compositions except insofar as constraints are imposed at invariant points, such that the number of degrees of freedom (F) is equal to zero. We explore the general case where F > 0.

An example of a solid with variable composition is C-S-H, the Ca/Si ratio of which can vary approximately between molar ratios of 1.8–0.8. In its coexistence with $Ca(OH)_2$, as occurs in normal portland cement pastes, C-S-H is calcium saturated (ratio ~1.8), but in blended cements, where $Ca(OH)_2$ is absent, this restraint is also absent and the Ca/Si of C-S-H may decrease significantly. The question arises of how to describe C-S-H of variable composition and define its coexistence and composition in thermodynamic terms. This is especially difficult as C-S-H may coexist with other solid phases also containing Ca, Si, and water, some or all of which may also be of variable composition. Experimentally, a number of procedures have been used to determine the partition of a given species or species amongst coexisting solids and an aqueous phase (e.g., tie lines, tie triangles, etc.).

However, systematic experimental observations of this kind are not often sought and recorded in the literature on cement science. One approach has been to arbitrarily define a series of C-S-H phase compositions each having a fixed Ca/Si ratio and creating a database, allowing the computer to select by iteration the C-S-H composition most appropriate to each step of the calculation. Another, more elegant, approach is to employ the Gibbs-Duhem equation, as was done by Kersten and colleagues,[16] relating the change in thermodynamic properties to changes in composition. Thus, several methodologies exist,[17-25] but ultimately the goal is to minimize the overall free energy of a heterogeneous system and thereby define the composition of coexisting phases. But no methodology, however elegant its mathematics, will overcome database deficiencies. Fortunately we have reasonable property-composition data for most of the main cement hydrate phases, including C-S-H, as will be presented subsequently.

Other relationships between free energy and equilibrium constants enable thermodynamic data to be derived from experimental data and vice versa. An important class of reaction involves oxidation and reduction. For example, sulfur in cement systems is normally present as sulfate, SO_4^{2-}; that is, sulfur is formally present as S(VI). But important exceptions occur. Iron blast furnace slag contains typically ~1% S, mainly as sulfide sulfur, S^{2-}. In the course of hydration, reaction ensues between the sulfur species present in slag with that in cement. Many reaction products are possible, but an important reaction product appears to be thiosulfate, $S_2O_3^{2-}$; thiosulfate formally contains sulfur in the S(IV) oxidation state. Reactions between sulfur in its various oxidation states require transfer of electrons as well as of oxygen, hydroxyl, and water, depending on the specific reaction product(s). However, once a balanced equation is written, we can take advantage of another method of relating the resulting free energy change to the energetics of electron transport reactions and use the relation:

$$\Delta G = - nFE \tag{25}$$

where n is the number of electrons requiring to be transferred per atom, F the Faraday constant (96 500 C/s), and E the potential (in volts) for the

---

*Usually defined as unit concentration (1 molal) for aqueous phases, 1 bar pressure for gases, and either 20 or 25°C. Where H^+ ions are involved, the concentration condition is equivalent to pH 0.

reaction. Extensive tabulations of E_o, the standard potentials for defined half reactions, one an oxidation and the other a reduction, are available in the literature. Like other free energy changes, these may be combined by addition, taking into account algebraic signs and conservation of the total numbers of electrons to determine an overall reaction potential. However, few reactions occur under standard conditions* with the result that E_o values may need to be transformed to the actual conditions. The necessary transformation equations are well known and are presented subsequently; many computer programs will automatically transform standard ΔG (or E_o) values to the relevant nonstandard conditions. We subsequently also explore the application of electrode potentials to steel embedded in cement paste.

Particle Size and Solubility

Cement systems often contain fine-grained hydrates and one frequently raised question is the impact of particle size on solubility and hence free energy. The relationship, given by the Ostwald-Freudlich equation is

$$S_r / S_i = \exp (2EVR^{-1}T^{-1}n^{-1}) \tag{26}$$

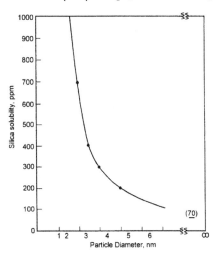

Figure 6. Solubility of amorphous silica as a function of particle size. The line at 70 ppm (right side) is the solubility for "infinite" size particles.

where S is the solubility of a particle of radius r and S_i is the solubility of a particle of infinite radius. The ratio is a function of E, the interfacial surface energy; V, the molar volume; R, the gas constant; and T, the temperature in Kelvins. While the equation finds use generally, values for E do not seem to be well known for many cement substances. However, if we use data for hydrated colloidal silica, which may well be representative of silicates (e.g., C-S-H), the effect of radius of curvature on solubility is readily apparent (Fig. 6). The nominal solubility for SiO_2 particles of infinite radius (shown by a dashed line in the figure) is ~70 ppm, but solubility increases as particle radius decreases, slowly at first but with the most significant increases occurring for particles <10 nm.

Particle morphology and the geometry of interfacial contacts also influence solubility. In systems with particles of different sizes, particles with positive radii of curvature tend to dissolve and accrete on other particles having larger (or infinite) radius of curvature. The lowered solubility in regions having negative radii favors agglomeration and growth at these sites. In a hydrated cement paste it is probable that the total mass of C-S-H (or other colloidal substances) in true solution is low after the first few hours or days of hydration. However, elevated solubilities, arising from the presence of small radius particles, are probably important in the first few minutes or hours of hydration and play a significant role in through-solution transport and precipitation processes during early hydration and in the structural organization of amorphous phases (e.g., C-S-H) at the nanoscale. We will return to this point subsequently.

Indirect evidence exists showing that C-S-H consists of an agglomeration of small subunits, or domains, each of colloidal dimensions; for example, see Ref. 26. These considerations are relevant to establish the true solubility of C-S-H since, as shown above, particle size and geometry affect solubility. Published data for C-S-H solubility exhibit wide scatter and it has always been supposed that the scatter reflects analytical error at least in part, especially for silica; even traces of hydrated colloidal silica persisting in the aqueous phase could elevate the analytical results. While these problems of analysis remain, it is possible to define with reasonable accuracy the properties of C-S-H.

The properties of well-aged C-S-H gels were established by Glasser et al.[27] from data in the literature. The free energy of formation, ΔG_{of}, of C-S-H was defined as:

$$\Delta G_{of} = [2(Ca/Si) - y]\, \Delta G_{of}\,(Ca^{2+}) + y\Delta G_{of}\,(CaOH^+)$$
$$+ 2\,G_{of}\,(H_2SiO_4^{2-})$$
$$+ [4(Ca/Si) - 4 - y]\, [G_{of}(OH^-) + RT\, lnK_{sp}] \qquad (27)$$

Auxillary plots of the distribution of silicate species and of K_{sp} as a function of pH are needed. The results are shown in Table IV. Note that ΔG_{of} varies lineraly as a function of Ca/Si ratio, as shown in Fig. 7. The data were in good agreement up to Ca/Si ratios ~1.4–1.6, but did not give good internal consistency at high Ca/Si ratios, leading to the suspicion that either

Table IV. Solubility and thermodynamic data for calcium silicate hydrogels at 25°C

Ca/Si	$[Ca^{2+}]_{total}$ (mmol·L⁻¹)	$[OH^-]$* (mmol·L⁻¹)	$[H_2SiO_4^{2-}]$ (mmol·L⁻¹)	$[H_3SiO_4^-]$ (mmol·L⁻¹)	I (mol·L⁻¹)	g'	K_{sp}	K'_{sp}	G_{of} (kJ·mol⁻¹)
1.2	2.1	3.2	6.9	1.53	0.023	0.059	8.08×10^{-16}	4.77×10^{-17}	-3920
1.2	1.3	2.3	0.12	0.35	0.022	0.063	9.05×10^{-18}	5.67×10^{-19}	-3930
1.2	4.4	8.3	0.047	0.043	0.022	0.066	8.27×10^{-17}	5.50×10^{-18}	-3920
1.2	6.1	12.0	0.03	0.020	0.022	0.067	1.04×10^{-16}	7.02×10^{-18}	-3920
1.2	7.3	13.8	0.031	0.019	0.022	0.067	1.67×10^{-16}	1.12×10^{-17}	-3920
1.2	8.5	16.5	0.013	0.007	0.020	0.078	5.14×10^{-17}	4.03×10^{-18}	-3920
0.93	2.13	2.51	0.070	0.181	0.039	0.055	2.83×10^{-13}	1.55×10^{-14}	-3440
1.01	3.90	5.89	0.043	0.047	0.038	0.049	2.14×10^{-14}	1.04×10^{-15}	-3580
1.08	5.28	8.32	0.035	0.027	0.039	0.042	3.35×10^{-15}	1.42×10^{-16}	-3710
1.17	4.71	7.94	0.036	0.029	0.039	0.034	1.81×10^{-16}	6.19×10^{-18}	-3870
1.25	5.72	8.91	0.025	0.018	0.039	0.029	1.35×10^{-17}	3.86×10^{-19}	-4020
1.33	7.42	12.60	0.019	0.010	0.039	0.024	2.46×10^{-18}	6.02×10^{-20}	-4160
1.43	6.45	12.60	0.022	0.011	0.039	0.020	1.46×10^{-19}	2.86×10^{-21}	-4340
1.49	9.42	13.80	0.015	0.007	0.039	0.017	3.48×10^{-20}	5.83×10^{-22}	-4450
1.57	10.50	14.50	0.017	0.008	0.040	0.014	1.55×10^{-20}	2.19×10^{-22}	-4590
1.63	15.00	17.80	0.013	0.005	0.035	0.016	9.55×10^{-21}	1.52×10^{-22}	-4700
0.97	2.0	3.84	0.039	0.072	0.034	0.059	1.70×10^{-14}	1.01×10^{-15}	-3510
1.07	3.0	5.90	0.028	0.035	0.034	0.049	7.45×10^{-16}	3.68×10^{-17}	-3690
1.18	4.0	7.94	0.020	0.019	0.034	0.039	2.61×10^{-17}	1.03×10^{-18}	-3890
1.25	5.0	9.95	0.015	0.012	0.034	0.034	3.42×10^{-18}	1.17×10^{-19}	-4020
1.32	6.0	12.00	0.012	0.008	0.034	0.030	5.03×10^{-19}	1.49×10^{-20}	-4150
1.46	8.0	16.00	0.007	0.004	0.034	0.022	1.04×10^{-20}	$2.2.7 \times 10^{-22}$	-4400
1.59	10.0	20.00	0.005	0.002	0.035	0.016	3.46×10^{-22}	5.71×10^{-24}	-4630

*$[OH^-]$ values are total hydroxyl concentrations except those for Greenberg and Chang, which were calculated from pH.

a structural change occurs at Ca/Si ~1.6 or that high Ca/Si ratios result from occlusion of $Ca(OH)_2$, or both.

However, Jennings has interpreted data differently, finding that different "varieties" of C-S-H exist, each having its own well-defined solubility curve.[28] These varieties are characterized by differences in crystallinity and in the nature of the crystalline organization, having either a tobermorite or jennitelike structure, or mixtures thereof.[29,30] This explanation does not inherently conflict with thermodynamics, but it neglects or subordinates the effect of particle size — in this case, the size of the units of agglomeration in C-S-H. As we have shown, solubility increases as the constituent particle size decreases. In structural terms this increase arises as a consequence of the difficulty of satisfying both valence and coordination requirements at or near solid surfaces. A geometric calculation suffices to show the extent of structural perturbation associated with particle size. If we assume that the near-surface, structurally disturbed region of a solid extends to a depth of 0.5 nm from the surface, and that the particles have spherical geometry, Fig. 8 shows the total volume fraction calculated to lie within the disturbed zone. As can be seen, the volume fraction of solid in the disturbed zone is small above ~100 nm, but increases rapidly,

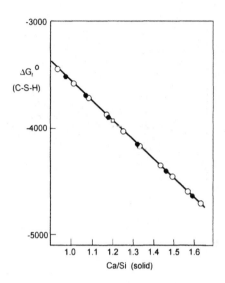

Figure 7. Variation in free energy of formation of C-S-H as a function of Ca:Si ratio at ~25°C. Symbols for the data points are taken from the literature and defined in Ref. 21.

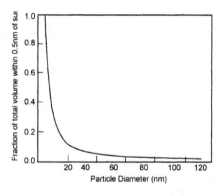

Figure 8. Impact of particle size on the volume fraction of near surface material, defined as within 0.5 nm of the geometric surface of spherical particles.

particularly as particle sizes decrease to less than 20–40 nm. The calculation could of course be repeated assuming other geometries and other disturbed depths but with similar results, showing that the solubility of nanoparticles is increasingly expected to deviate from those of micro- and macroparticles. The effect on solubility is reduced but not eliminated in C-S-H by flocculation; because of poor flocculation, as evidenced by the characteristic occlusion within C-S-H of nanopores, much internal surface is created with small radii of curvature and hence with elevated solubility. The impact on solubility should in theory be continuous; Eq. 26 does not predict discontinuities. Thus the size distributions of the flocculant units should give rise to an infinite number of solubility curves, rather than on discrete curves as reported by Jennings.

Moreover, as the size of the units of agglomeration decrease, particularly below ~100 mm, crystallinity must inevitably decrease owing to the structural disturbances associated with surfaces. But in larger units of agglomeration (approximately >100 nm) a range in crystallinity could occur. Correspondingly at >100 nm, an infinite number of order-disorder states could exist within the subunits comprising agglomerates, leading to a corresponding infinite number of solubilities. These solubilities would however, be constrained to lie within maximum and minimum values: for a fixed unit size, the former would correspond to the most disordered state and the latter to the ideal "crystalline" state.

It thus appears that C-S-H solubility depends upon two external variables — temperature and composition — as well as upon two internal variables: domain size and order-disorder state. At a fixed temperature and composition (Ca/Si ratio), the solubility is envisaged as being controlled by a combination of the size of the subunits of agglomerates and crystallinity. This state is probably achieved or closely approximated by aged C-S-H. However, as the agglomerated size of subunits in C-S-H decreases to approximately <100 nm, crystallinity cannot be high and thus flocculate size becomes the principal control on solubility. This state is probably best exhibited by fresh, newly flocculated C-S-H. Figure 9, containing unpublished data by Hong,[31] shows the influence of composition and crystallinity: solid diamonds and open squares show the solubility trends, as reflected by the pH conditioning ability of a series of gels prepared and aged at 20°C. The same gels (open squares) were also aged at 55 and 85°C without crystallization occurring in the normal sense of the term; nevertheless, solubilities at 25°C move to lowered pH conditioning of the aqueous phase as the cure temperature increased. Finally, treatment at 180°C, shown by open cir-

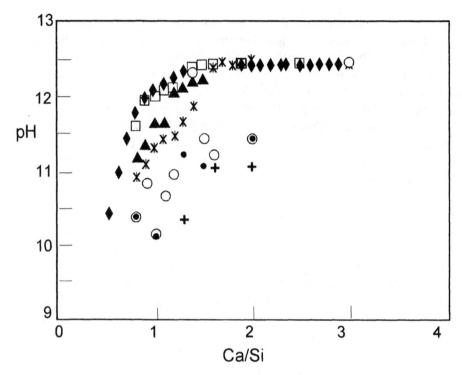

Figure 9. pH conditioning ability of C-S-H gels from Hong[41] and from the literature (♦). The gels were treated without crystallization at 55°C (Δ) and 85°C (*) and with crystallization or partial crystallization at 180°C (O). Points designated "+" refer to (from left to right) pure synthetic 11 Å tobermorite, xonotlite, foshagite, aflwillite, and hillebrandite.

cles, led to crystallization. Each set of conditions relevant to gels gives a smooth trend line; with increasing severity of treatment the curve moves to lower pH. It is predicted that an infinite number of curves could thus be generated by appropriate treatments, although this cannot be proven experimentally. The data generated comprise a series of snapshots with time as an additional variable. However, the 180°C points form an irregular trend; a different set of rules governs the pH conditioning ability of crystalline phases and phase assemblages. For example, some of the crystalline phases obtained at 180°C are actually metastable at ~20°C and, on that account, tend to give generally higher solubilities than phases stable at 20°C. Moreover, data points cannot simply be joined for mixtures of crystalline phases; phase rule conditions and restrictions apply when two phases equilibrate with an aqueous phase.

The family of solubility curves characteristic of accessible structural states thus have upper and lower limits. These limits, and the differences between them, have to be found by experiment. However, application of thermo-dynamic principles assists in the interpretation of data.

C-S-H obtained at ~25°C consists of an assemblage of nanoparticles that incompletely fill space. This packing gives rise to nanopores, which are a characteristic feature of C-S-H. The volume of nanopores determined, for example, by N_2 soprtion is relatively large, comprising perhaps 10% of the total volume. Relatively little direct evidence emerges concerning how interfacial effects affect the properties of liquid nanopore contents, but it is reasonable to assume from the large depression of the freezing point of water in nanopores, observed by cryogenic calorimetry, that the properties of "water" in nanopores differ significantly from those of bulk water.

It is apparent from the foregoing that the thermodynamic properties of C-S-H must be subject to considerable further refinement. Most work has been done on solubility modeling and relating solubility to internal crystallinity and to the specific crystallite structures, either tobermorite or jennite. Other materials of low dimensionality (e.g., organic high polymers and "amorphous" metals) are known; while not directly analogous to C-S-H (in structure), similarities do exist. Thermodynamic approaches to elucidation of the properties of these diverse materials may serve as a link to explain the properties of C-S-H and develop constitutional models.

Osmotic Pressures

In the course of cement hydration, particularly early hydration, it is readily apparent that substantial local differences in composition of the separate solid grains occur and that local differences in the composition of aqueous solution could develop spontaneously as reaction progresses. Of course, the direction of spontaneous changes is such that it evens out differences in composition of the aqueous phase; the solids of course remain polyphasic. But this homogenization takes time and requires ion transport to occur. In the meantime, differences in aqueous concentration may, in the presence of ion-selective membranes, lead to spontaneous development of internal pressure gradients, termed "osmotic pressure."

Figure 10 depicts a simple experiment designed to show the origin of these osmotic gradients and their expression as pressure differences. In the example, an inverted funnel has been covered by a membrane and immersed in water. The space above the membrane is filled with a solution of a nondiffusing species (e.g., sugar) and the liquid levels are adjusted so

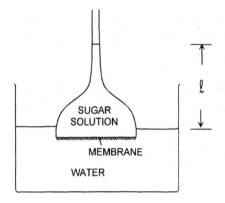

Figure 10. Classical demonstration of osmotic pressure in which a sugar solution is placed in an open bell that is in turn contacted with distilled water. Initially the liquid levels are made equal. However, water diffuses through the membrane, creating the difference in liquid levels. The pressure exerted at steady state by this head is proportional to l, the head. This equilibrium pressure is known as the osmotic pressure of the solution.

that at the outset the two liquid levels, inside and outside, are equal. As time elapses, the liquid level in the sugar solution will be seen to rise; the pressure equivalent to the rise (l) is termed the "osmotic pressure" (O_p) such that

$$O_p = kcT \tag{28}$$

where C is the ratio of moles of solute to volume of solvent, T is the temperature, and k is a constant so that

$$O_p = nkT \tag{29}$$

This may be compared with the equation of state for ideal gases, which takes the very similar form:

$$PV = nRT \tag{30}$$

where P is pressure, V is volume, n is the number of moles, T is the temperature in Kelvin, and R is the gas constant, 8.314 J/mol. Experimental measurement of the osmotic pressure exerted by solutions discloses that the numerical value of k in Eq. 29 is also close to 8.314 J/degree.

In atomistic terms, an osmotic membrane must be permeable to water molecules but not to solute ions or molecules. Water in the funnel is effectively diluted by sugar, the solute, and to achieve an equal water activity on either side of the membrane, water must diffuse inward. As a consequence, the liquid level rises in the neck of the funnel. However, the increasing pressure of the column of water in the funnel slows the rate of transport of water until a pressure balance is reached. The excess pressure achieved by the column of water at the balance point (i.e., the point at which the rates of

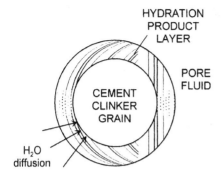

HYDRATION PRODUCT LAYER

PORE FLUID

CEMENT CLINKER GRAIN

H₂O diffusion

Figure 11. Supposed origin of osmotic pressure in a hydrating cement paste. As clinker hydrates, an envelope of hydration product develops. This envelope acts as a membrane. Ions dissolving at the interface between the clinker and the hydration product give rise to local supersaturation, as a result of which water diffuses inward through the hydrate layer, thereby generating an osmotic pressure.

migration of water across the membrane become equal) is termed the osmotic pressure. Even quite small concentration differences can thus achieve substantial pressures, perhaps equivalent to several atmospheres.

Osmosis has been suggested to be important in the early hydration of cement.[32] As is well known a period of dormancy occurs in early hydration (see Fig. 1). It is suggested that a shell of initial hydration products develops around clinker grains, as shown in Fig. 11. This shell retards access of water to anhydrous clinker materials and initiates a dormant period. However, the clinker grains continue to furnish ions to solution, albeit slowly. Since dissolution occurs selectively at the base of the membrane (i.e., the surface in contact with clinker), the solution becomes more concentrated than the mix water. As a result, water diffuses inward from the less concentrated mix water, leading to creation of an osmotic pressure across the membrane. This pressure eventually physically ruptures the membrane, leading to resumption of rapid hydration.

While the theory successfully accounts for the dormant period and its eventual termination, it has problems. One such problem lies in proving that the membrane has the necessary properties to sustain osmosis. It must be permeable to water but not permeable, or very much less permeable, to the dissolved hydrated cations, mainly Na, K, Ca, and so forth, necessary to sustain an osmotic pressure. Nevertheless, the concept is theoretically tenable although it remains to be proven that the early hydration products actually possess the ability to act as an osmotic membrane. However, membrane formation is assisted by high silica activities and it is likely that substances such as slag and fly ash have better potential for membrane formation than portland cement grains. Osmosis may also be an important source of disruptive pressures in mortars or concretes made with reactive siliceous

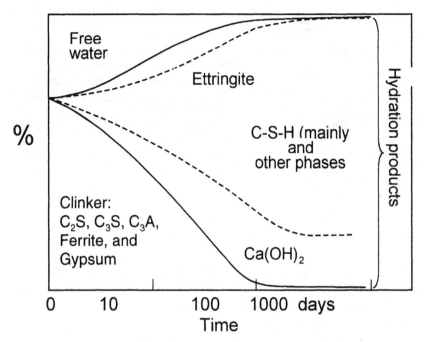

Figure 12. Schematic diagram showing the evolution of cement paste mineralogy as a function of time. The assumption is made that the amount of water is exactly that required to achieve full hydration.

aggregates.

Interactions between Equilibria and Kinetics

Most processes occurring in cements are mediated by kinetics; the kinetic factors are too numerous to list and are, in any event, process sensitive. In some cases we do not have the necessary kinetic data with which to undertake calculations. However, even where data are missing, application of a few principles will greatly assist understanding and perhaps permit semi-quantitative calculations to be undertaken. Two examples will be pursued, one involving the concept of reactive fraction and the other using the concept of local equilibria.

The concept of reactive fraction is best developed with the aid of a diagram (Fig. 12). The left side of Fig. 12 depicts the phase composition at the onset of hydration. For simplicity, it is assumed that the amount of water added is just sufficient to hydrate the cement so that as cement hydration approaches 100%, free water vanishes. Hydration products are only partly differentiated in the diagram. As shown earlier, cement clinker behaves uniformly once the very early stages (0–48 h) of hydration are past so that C_3S, C_2S, and C_3A — but not necessarily ferrite — hydrate at effectively the same rate. The bulk ratios of Ca/Al/Si in the hydrate products do not therefore differ much from those of the clinker (iron and sulfur possibly excepted) and remain approximately constant as a function of time.

But an important exception does have to be made for sulfur, initially present in cement mainly as sulfates ($CaSO_4 \cdot 2H_2O$, K_2SO_4, and Na_2SO_4), calcium langbeinite, or as mixtures of sulfates, for the following reasons. Much of this sulfate dissolves rapidly, within the first few minutes to days of hydration. Thus the ratio SO_3/Σ (other hydrate components) in the paste will be effectively high at the outset of hydration but decrease with time; as hydration progresses, the sulfate becomes effectively diluted in the reacted fraction. However the paste mineralogy is quite sensitive to the numerical value of this ratio; while the C-S-H and $Ca(OH)_2$ phases may not be directly affected, sulfate availability is important in determining the nature of sulfoaluminate precipitating at any given point in time. With decreasing sulfate availability, the progress of aluminate phase formation mirrors this availability; pore fluid sulfate concentrations decrease and the order of preferential solid phase formation at ~20°C is typically observed to be ettringite → sulfate AFm → mixed hydroxy-sulfate AFms. These ranges of preferential phase formation often overlap, and the transition between regimes occurs gradually. Of course the phase or phases being precipitated also differ in calcium content, so their formation or resorption indirectly affect the amounts of $Ca(OH)_2$ and C-S-H, as well as the amount of sulfate sorbed into C-S-H.

If we were to model these processes, we would first have to identify the composition and distribution of sulfate and the time at which it becomes available for reaction, and in what amounts, to obtain a series of snapshots showing the course of phase development with time. Thus kinetics mediate phase development. Conceptually, we would need to begin by expressing the change in chemistry of the reactive parts of the system as a function of time. Although the task appears daunting at first sight, and an infinite number of models might in theory be developed, two simple robust models are readily developed assuming either (1) fast and total release of sulfate within the first few minutes or hours of hydration or (2) slow release, in which sulfate is released at the same rate as Ca, Al, Si, and so on. In the case of fast release the ratio of SO_3/Σ (other components) would decrease with time, while in the second case it would remain constant. The latter assumption is, in any event, a precondition for calculation of the equilibrium reaction path. It is probable that most commercial cements are intermediate in behavior between fast and slow sulfate release. Depending on the nature of assumptions made, of course influenced by available experimental data and observations, the composition of the reactive fraction, and hence the reaction path, could be defined and then used to estimate mineral stabilities and amounts of respective phases, using a scheme akin to a Bogue calculation but modified to include hydrates. While we do not at present have sufficient data to implement this model, in principle it would not be difficult to do so, and a computer-based model with variable inputs would be helpful to explore how paste mineralogy changes in response to sulfate availability. This type of model, termed "reaction path" modeling, is commonly used in geochemical studies; numerous validated codes exist that, in conjunction with an appropriate database, could be applied to reaction path modeling of the impact of hydration kinetics on paste mineralogy. Thus, while we have a potentially infinite number of reaction pathways, the overall difference in free energy, assuming the same set of reactants and products, is independent of pathway, and although an infinite number of pathways could in theory occur, in practice only a few usually prove to be important.

The second example concerns partial replacement of cement by slag. In general, slag hydrates more slowly than cement, so, at early and intermediate stages of reaction, the reactive composition of the paste fraction reflects to a greater extent the composition of the cement than of slag. Exact calculations depend on assuming (or determining from experimental inputs) the chemical composition of slag and cement and the specific fraction of each that has reacted as a function of time. In general, iron blast furnace slags contain more silica, magnesia, and alumina than cement but less iron oxide, so the outline of how the changing chemistry of the reactive fraction changes is apparent. Owing to differential reaction rates, it is also clear that greater mineralogical changes will occur in the course of hydration of a blended cement than in a system formulated only with cement: this conclusion is attributable to the large differences in chemical composition between slag and cement as well as their different rates of reaction with water and with each other.

Calculation of the final steady-state phase distribution (amount, mineralogical nature) of a 70% portland cement–30% slag cement were undertaken using a generic model.[33] The model calculated six possible solid

Table V. Phase assemblages developed in cement–fly ash blends

With $Ca(OH)_2$:

        C-S-H (Ca/Si ~1.7), C_3AH_6, AFt, hydrotalcite, H_2O

$Ca(OH)_2$ with C-S-H (only Ca/Si ratio shown):

        1.7–1.5, C_3AH_6-$C_3AS_{0.43}H_{5.14}$, AFt, HT, H_2O

        1.5, $C_3AS_{0.43}H_{5.14}$-$C_3AS_{0.76}H_{4.48}$, AFt, HT, H_2O

        1.5–1.2, $C_3AS\bar{H}_{0.76}$, C_3ASH_4, C_2ASH_8, AFt, HT, H_2O

        1.2, C_3ASH_4, C_2ASH_8, AFt, H_2O

        1.2–0.85, C_2ASH_8, AFt, HT, H_2O

Notes: Owing to the low Mg content of class F fly ash, the amounts of HT (hydrotalcite) are small. By H_2O is meant a pore fluid saturated with respect to the relevant solids.

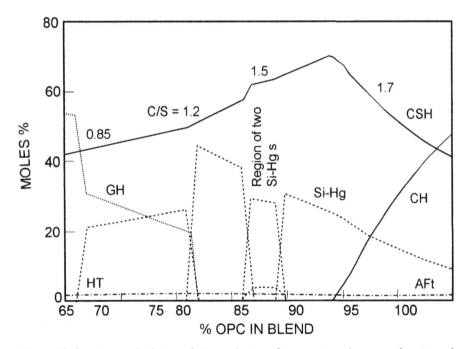

Figure 13. Specimen calculation of the evolution of paste mineralogy as a function of blend proportions, in this instance of a 70:30 blend of cement and slag.

phase assemblages, shown in Table V. Results of the calculation for the blended cements are shown in Fig. 13. Note that hydrogarnet was approximated by three solid solution compositions in the series C_3AH_6-C_3AS_3; in real-life situations, the three separate hydrogarnet compositions could be added together, although with the reservation that at least one solid miscibility gap may occur in the hydrogarnet series.[34] Calculation shows how CH is rapidly reduced in amount by adding slag. The rate of reduction in CH content is, however, rather sensitive to slag composition. As CH reacts, it is replaced by hydrogarnet, a hydrotalcitelike phase, and gehlenite hydrate. In the specimen calculation, the amount of C-S-H was found to reach a maximum at ~85–90% OPC content. At lower OPC contents, both the amount of C-S-H and its Ca/Si ratio decreased. The main advantage of calculation is speed and flexibility. While experimental studies on these blends required more than 5 years and were able to embrace only one slag and one cement composition in mix proportions at fairly coarse composition intervals (5–10%), calculations are rapid and could, if desired, be repeated for any cement and slag composition. Thus calculations of mineralogical evolution, supported and interfaced with experimentation, are desirable even in their present state of development. It is noteworthy that with the exception of C-S-H, these models conform to the equilibrium phase distribution.

In parallel with calculations, observations on the microstructure of maturing slag-cement pastes reveal a sharp spatial distribution of the solid products of reaction; phases characteristic of slag hydration such as gehlenite hydrate, or stråtlingite, a silica substituted AFm, as well as phases containing essential magnesium including the hydrotalcitelike phase and M-S-H gel, are often found to be physically localized within the outlines of former slag grains.[35,36] This suggests that magnesium is relatively immobile. Whether these minerals have reached phase equilibria is unclear; they probably have reached, or nearly reached, phase equilibria after 5 years of moist cure duration but have almost certainly not yet reached textural and microstructural equilibria.

Oxidation and Reduction (Redox) Processes

Several ancillary issues arise concerning slag cement blends. Sulfur is normally present in about the same amounts (1–3%) in both slag and cement. But the initial speciation of sulfur in slags is S^{2-} (sulfide) whereas it is sulfate (SO_4^{2-}) in cements. The final speciation of sulfur is a function of pH and Eh; the latter is a measure of the oxidizing (or reducing) potential of a solution. As noted previously, in alkaline solution and in the absence of oxygen (gas), sulfide and sulfate react to form thiosulfate. The conventional balance in alkaline solution is:

$$6H_2O + S^{2-} + 3(SO_4)^{2-} = 2(S_2O_3)^{2-} + 12OH^- \qquad (31)$$

aqueous   slag   cement   aq. +solids   aqueous

The source of the reactants and distribution of products is indicated. The product of reaction between sulfur species, thiosulfate, is partially soluble and can be detected in pore water by ion chromatography. Thus a new species appears, not anticipated by conventional cement chemistry but predicted to occur by thermodynamics and verified by experiment. As a consequence, another important physiochemical parameter emerges: the internal oxidation-reduction (redox) potential of the matrix and its associated pore fluid.[37,38]

The internal oxidization-reduction potential of cements is measured on a scale relative to the standard hydrogen electrode, the voltage for which is arbitrarily taken as zero. Normally, portland cement is produced under oxidizing conditions and, as a consequence, has an oxidizing potential of +100 to +200 mV. However, in the presence of reduced sulfur species (S^{2-}, $S_2O_3^{2-}$, etc.), this potential diminishes to –200 to –400 mV; i.e., the internal cement environment becomes reducing. Moreover, while the oxidization-reduction potential of plain portland cement is essentially unbuffered, that of slag cement is buffered by the presence of both reduced and oxidized forms of the same couple. Technically, oxidation-reduction couples producing this buffering action are termed "poised" and the capacity of the system to resist change is termed "poising capacity," but this author applies the more familiar term "buffered"; however, the concept underlying both usages is essentially identical.

Many of the commonly encountered reactions occurring in the course of cement hydration do not involve oxidation and reduction reactions; for example, calcium, magnesium, aluminium, silicon, sodium, potassium, hydrogen, and oxygen do not normally change in oxidation state in the course of cement hydration. On the other hand, the color of hydrated cement paste is associated in part with manganese, which exhibits variable oxidation states and is affected by pH and Eh. But perhaps the best-known and most important phenomena occurring in cement and concrete concern the use of embedded steel as reinforcement. The thermodynamics are straightforward: iron (including in this context ordinary mild steel, containing ~0.3 wt% carbon) is thermodynamically unstable in the cement environment. After any dissolved oxygen is consumed by oxidation of steel — a relatively rapid process — reaction between steel and concrete in a closed system must proceed with liberation of hydrogen. The nature of the solid iron corrosion product depends on oxygen availability. At low availability, $Fe(OH)_2$ or a compound such as "green rust" is obtained by reaction of Fe(II) with water and cement components. At higher oxygen availability, Fe_3O_4-γFe_2O_3 — a spinel structured solid solution — or one or more iron (III) hydroxides are obtained. The iron corrosion reactions requiring H_2 liberation are known to be slow at normal service temperatures, and, moreover, the products of reaction are rather insoluble at high pH and form a semi-protective layer on steel, thus kinetically limiting ingress of water necessary to sustain reaction. This condition of slow steel corrosion is associated with the passive state; corrosion rates on the order of 1 mm/year are often reported.[39] In this condition, a zone of low redox potentials, perhaps as low as –400 to –500 mV, develops in the vicinity of the steel-cement paste interface. On the other hand, the service environment controls the oxygen potential at or near the concrete surface; in air it is ~0.21 atm. Thus substantial electrochemical gradients in oxidation potential may develop spontaneously. Steel, cement paste, and air are thermodynamically incompatible.

Electrochemical data relate closely to free energy changes; for example, the free energy change was related to electrochemical potential in Eq. 25.

A brief thermodynamic appraisal of the stability of steels in cement is as follows. In good-quality concrete, such that oxygen permeability is low, embedded steel consumes oxygen locally with the result that potential (relative to the standard hydrogen electrode) decreases to ~–500 mV. Thereafter, and in the absence of oxygen ingress, corrosion continues, albeit slowly, by H_2 evolution by reaction of steel with pore fluid. The stable corrosion film formed on substrate steel adheres on an atomic level; the densely packed structure of spinel permits only slow diffusion of protons (which, in any event, are scarce in high-pH pore fluids) with the result that corrosion is also slow. Since reinforcing steel is generally of low quality, physical inclusions and inhomogenities in the metal tend to produce defects in the cover oxide coat and, not surprisingly, localized or pitting corrosion often becomes superimposed on more general corrosion. However, pitting corrosion may also arise for other reasons.[40]

The mineralogical nature of the protective coating is affected by the presence and concentration of a number of species, particularly chloride. At high pH, the stability of the spinel-structured passivating layer is disrupted. For example, the akaganeite phase, β-FeO(OH,Cl) is stabilized with respect to spinel even at low chloride contents. Its tunnel structure is open to ion migration and does not afford as good of a self-protecting layer as spinel. As far as the author is aware, no systematic experimental or thermodynamic studies have been made of mineral stabilities in the relevant portions of the Fe-O-Cl-H$_2$O system. But if this were to be done and coupled to the electrochemical literature, understanding of the equilibria and kinetics of steel corrosion in the presence of chloride both in fresh and aged cement and in blended cement pastes might be greatly advanced.

Electrochemistry is thus a direct link between thermodynamics and cement science. It has special appeal because of the ease of measuring accurately low voltages. To facilitate measurement, the two half-reactions, one an oxidation and the other a reduction, have to be physically separated and reaction enabled to proceed by constraining electrons to flow through an external circuit. The two half-reactions are defined by the anode, at which oxidation occurs, and the cathode, at which reduction occurs. In the example developed in the text, the corrosion of steel embedded in cement paste (or concrete; the aggregate is generally neutral), the embedded steel becomes anodic and the region in which oxygen is chemically reduced to oxide, O^{2-}, or OH^- ions, comprises the cathode. The usual source of oxygen is the cement/air interface, thus, in real-life corrosion, physical separation between anode and cathode may occur spontaneously. This spontaneous separation of anode and cathode is observed in many other practical instances of corrosion.[40] But if the spontaneous separation is not sufficient for physical measurements, a little ingenuity is usually sufficient to achieve the necessary separation in laboratory simulations.

A link between kinetics and equilibrium is also readily achieved in the course of electrochemical measurements. In a simple example, oxidation of iron to ferrous (FeII) iron,

$$Fe^0 + 2e^- = Fe^{2+}$$

Two moles of electrons must be transported per mole of iron (55.5 g) oxidized. The flow of current required for the oxidation of one mole is therefore $2 \times 96\,500\,C = 193 \times 10^3$ ampere seconds.

Other Application Areas

Phase Equilibria in the CaO-SiO$_2$-H$_2$O System

Approximately 20 mineral phases are known to occur in the CaO-SiO$_2$-H$_2$O system. The absolute thermodynamic stability of most of these solids is unknown but, taking into account the geological setting of those phases occurring as natural minerals, many are apparently stable only at high pres-

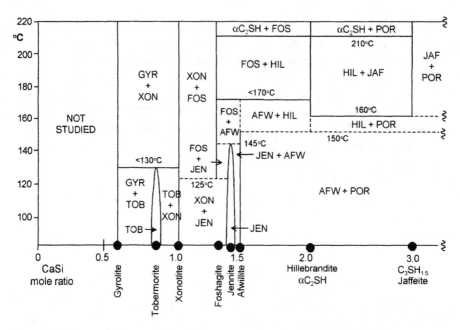

Figure 14. Phase relations at high calcium contents in the CaO-SiO$_2$-H$_2$O system up to 220°C.

sures, greatly in excess of those encountered even in autoclaved cements. The greatest technical interest is in those minerals occurring as stable phases at low or moderate pressures, typically not exceeding saturated steam pressure and at temperatures in the range 100–200°C, typically used in steam curing or in autoclaved cements. Much mineralogical synthesis work has been done, but previous investigators have often assumed that it was difficult, if not impossible, to reach true phase equilibria in this range of compositions, pressures, and temperatures. Consequently, thermodynamic data for crystalline C-S-H phases are of variable quality or lacking entirely, and most phase diagrams record mode-of-occurrence of minerals but without attempting to delineate fields of stability. However, the system has recently been revisited by Hong and Glasser.[41] They found that, contrary to impressions presented in the literature, it was possible to achieve reversible

---

*Strictly speaking, data for the system presented in Fig. 14 are not isobaric. However, the water vapor pressure at 200°C is modest, so the system is treated as being effectively isobaric and the discussion is related only to the impact of changes in composition and temperature.

reaction leading to formation or decomposition of a particular mineral phase within a few weeks or months. Figure 14 presents the thermodynamic relationships at saturated steam pressures in the form of a phase diagram. In the range of compositions studied (Ca/Si mole ratios ≥ 0.50), nine crystalline phases were encountered. Rather remarkably, only two of the nine solid phases — gyrolite and portlandite — are stable across the entire temperature range studied (90–220°C). Many systems have hydrate phases with minimum or maximum limits of thermal stability,* but it is unusual to find so many phases — seven in total — fitting this condition, and that also cluster in a relatively narrow range of temperature, pressure, and composition.

With respect to composition, all synthetic phases agree with the mineralogical and chemical data in the literature with the exception of jennite, whose synthesis ratio, confirmed by analytical electron microscopy, was close to Ca/Si = 1.45. This contrasts with the structure determination made on naturally occurring crystals, giving Ca/Si = 1.5. Both jennite and tobermorite are defect structures and probably have a short range of Ca/Si ratios, so that departures from ideal stoichiometry are not unexpected.

Data on the solubility of calcium-rich C-S-H phases are compiled in Ref. 42; the data were obtained at 25°C for well-characterized synthetics. Natural minerals were included in the experimental solubility determinations but data thus obtained were deemed unreliable owing to large releases to the aqueous phase of structurally nonessential elements (e.g., Na, K, Mg, and Sr) in the course of equilibration. The data presented in Table III have the advantage of having been obtained in the same program, using a consistent methodology. The greater thermodynamic stability of crystalline phases relative to C-S-H gels is demonstrated in Fig. 15, showing 25°C solubilities of both the crystalline synthetic phases and gels. Restricting comparison to equal Ca/Si ratios, the crystalline phases tend to have order-of-magnitude lower calcium solubilities than gel. Silica solubilities, although not shown, conform to this trend. These solubilities are reflected in pH conditioning ability: a gel conditions aqueous pH values typically 1–2 units higher than a crystalline compound of the same Ca/Si ratio.

Restricting comparisons to the same Ca/Si ratio, the lower solubility of crystalline compounds relative to gel shows that gel must be metastable with respect to a crystalline phase or phases. In describing gels in the literature, it is commonplace to describe some gel states as "stable." While the state thus described may be persistent and not change with time, and, indeed, to have a lower free energy than another, more disordered gel, a gel

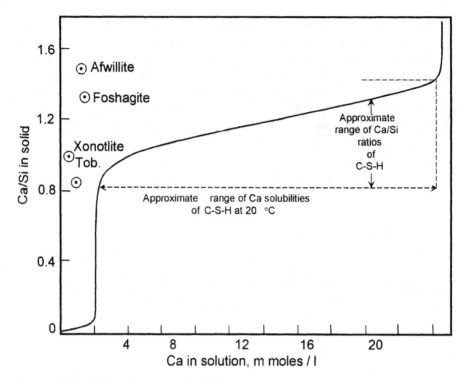

Figure 15. Solubility data in the CaO-SiO₂-H₂O system. The continuous curve is for C-S-H and its mixtures with either silica gel (low Ca/Si ratios) or Ca(OH)₂ (high ratios). Solubility data for selected crystalline C-S-H phases are also shown. Data obtained at temperatures in the range 18–25°C are used. "Tob" stands for tobermorite.

cannot be stable in the absolute thermodynamic sense. Therefore, discussions of stable or metastable gel states should always be qualified to ensure that the term "stability" is being used in the relative, not the absolute, sense.

Free energy of formation data are not well established at elevated temperatures for the CaO-SiO₂-H₂O crystalline phases, but the data do permit some internal checks. For example, tobermorite has an upper stability limit of 130°C, at which point it decomposes according to the reaction:

$$11 \text{ Å tobermorite} = \text{gyrolite} + \text{xonotlite} \tag{32}$$

It therefore follows that, with inclusion of stoichiometric coefficients, $\Delta G_{reaction}$ must be zero for the reaction at 130°C; if ΔG were known for any two of the solids, the third could be calculated, taking stoichiometric coefficients and changes in water content (if any) into account. Other examples of internal checks can be found by reference to Fig. 14. Most C-S-H mineral phases dissolve incongruently at ~25°C and, moreover, experimental problems in determining equilibrium solubilities preclude obtaining other than provisional free energy of formation data.

An important question arises concerning the application of these data to industrial practice; given that tobermorite is normally synthesized from reactants at 160–180°C in the course of autoclaving cement-silica mixes, why should tobermorite form outside its field of stability? The answer has to be given in two parts. First, the time allowed for tobermorite to form in industrial synthesis is typically brief, 12–18 h. If, however, the autoclaving time is extended, perhaps with a view to improving the yield of tobermorite, this is invariably unsuccessful: tobermorite decomposes to gyrolite and xonotlite, suggesting that its initial formation at 160–180°C occurred metastably. Second, the tobermorite made commercially is often anomalous tobermorite; this tobermorite variant is believed to be stabilized by solid solution of sodium and aluminium.[43] On account of differences in structure, composition, and defect contents between anomalous and normal tobermorites, both 11 and 14 Å varieties, the thermal stability of the anomalous form may increase. Thus it is possible that the anomalous phase is stable to higher temperatures than normal tobermorite. As our knowledge of the thermodynamic properties of the C-S-H phases, particularly tobermorite, expands, we may be able to fine-tune calculations made in support of synthesis using quantities appropriate for particular tobermorite structural variants and their solid solutions.

In summary, thermodynamic treatments can account for all solid C-S-H phases and their coexisting aqueous phases. On account of its importance to cements, much attention is paid to the C-S-H cement gel phase. Although metastable, it is persistent at ~20°C although it crystallizes rapidly at autoclaving temperatures (>100°C) and normally at 160–220°C. The finer details of C-S-H structure are currently the subject of much research. Although the extent and nature of the structural organization cannot be quantified by presently available techniques, it is apparent that the internal structure relates both to the physical size of agglomerate units as well as nanoscale ordering and, of course, Ca/Si ratio. Tobermoritelike and jennitelike structural units have been considered to exist in C-S-H. However, other structural models based on stable phases in the appropriate composition range (e.g., gyrolite, xonotlite, foshagite, and α-C_2SH) appear arbitrarily to have been excluded from the range of potential structural models. An advantage of a purely thermodynamic treatment is that it need not be linked to specific structures or constitutional formulae, although it is only fair to note that most thermodynamic models do, in fact, attempt correlations of this kind. At present, thermodynamic models of C-S-H reproduce well their input data — a first and necessary test of consistency — and usefully enable C-S-H compositions and coexistence, albeit metastable coexistence, with other phases to be explored. The success of next-generation models rests in part on obtaining more accurate input data, especially on the fractionation into tobermorite of minor components (e.g., Na, K, Mg, S) as well as the impact on Al and on the thermodynamic properties of C-S-H, and the phase relations of C-S-H to other phases, especially those characteristic of blended cements. On account of difficulties in quantification of the nature and amount of phases present in cement paste, even by the most advanced methods, thermodynamic models are attractive. Such models may not be absolutely reliable in their present state of development but, like a Bogue calculation for clinker, provide a tool with which to relate formulation, degree of reaction, and matrix mineralogy. This, in turn, will provide a platform to develop correlations with engineering properties.

Thermodynamic Analysis of the Thaumasite Form of Sulfate Attack

The discovery of thaumasite in degraded concrete exposed to sulfate attack is not new but has achieved prominence as more new cases are discovered. A recent symposium reviews and documents these occurrences.[44] Thaumasite — $Ca_6Si_2(SO_4)_2(OH)_{12}(CO_3)_2 \cdot 24H_2O$ — is structurally related to ettringite, but differs significantly in chemistry: Al in ettringite is replaced by Si, and, moreover, carbonate is an essential constituent. Despite chemical differences between ettringite and thaumasite, the unit cell sizes of the two are very similar; hence distinction by X-ray diffraction is difficult. Moreover, ettringite and thaumasite form a considerable range of mutual solid solutions.

The properties, synthesis, and stability of thaumasite have been the subject of different views in the literature.[45–47] What seems generally agreed is that (1) formation of thaumasite is relatively slow, requiring weeks or months to obtain good yields, (2) ettringite frequently appears in synthesis batches prior to thaumasite, leading to the supposition that ettringite is an essential intermediate in the formation of thaumasite, and (3) thaumasite is not stable at elevated temperatures (most successful synthesis have been done at <15°C).

Important questions for producers and users of concrete are: How can thaumasite sulfate attack (TSA) be avoided? What, if any, precautions can be taken to avoid deterioration of apparently well-designed and well-made concrete? To help answer these questions, Damidot and colleagues undertook thermodynamic calculations of the stable phase assemblages of thaumasite.[48] This was made possible by the availability of preliminary data on the thermodynamic properties of synthetic thaumasite. These can be computer assembled and checked to ensure that composition space is filled, that is, every chemical composition can be recast to a mineralogy. Calculation disclosed that 331 stable phase assemblages, including an aqueous phase, link the 15 known solids. Phase assemblages including thaumasite are shown in Table VI. The calculations for 25°C disclose that thaumasite is thermodynamically stable at this temperature; its thermal stability extends to higher temperatures than were hitherto supposed. In a number of assemblages (nos. 17, 20, 27, and 29), thaumasite can form directly from aqueous solution, and it is not essential to have ettringite as a precursor in order to obtain thaumasite. Nevertheless ettringite and thaumasite do coexist in a number of assemblages: those numbered 18, 19, 21–26, 28, and 30, perhaps giving the impression that ettringite is a precursor to thaumasite.

Table VI ranks the invariant points in order of increasing aqueous sulfate contents. From this ranking it can be seen that thaumasite requires a minimum aqueous sulfate concentration of 2.49×10^{-5} mol/kg in order to achieve stability. Other assemblages (nos. 1–16, not shown in Table VI, but see Ref. 48 for a complete list) have lower aqueous sulfate concentrations and as a consequence do not contain thaumasite. Within the range of sulfate contents of the aqueous phase necessary to stabilize thaumasite, we can distinguish two main patterns of coexistence of thaumasite with other phases: thaumasite in assemblages with and without gypsum. Gypsum becomes stable and coexists with thaumasite only at relatively high aqueous sulfate concentrations, $>1.24 \times 10^{-2}$ mol/kg.

Thaumasite is stable over a wide range of aqueous pH values. These range from a low of 8.51 (assemblage 29) to as high as 12.52 (assemblage 21). But the accuracy of calculating pH is probably +0.05–0.1, so other high-pH assemblages, those numbered 19, 24, 25, and 26, have essentially the same high pH, ~12.5. This means that even fresh and relatively unleached and uncarbonated cements have the potential to develop thaumasite, although clearly, if significant quantities of thaumasite are to develop, carbonate and sulfate must be present in appropriate amounts.

Application of these data to real cement systems can be made as follows. Let us initially consider concrete as a closed system — that is, its composition is unaffected except by exchanges of heat; it does not react or has not

Table VI. Composition of the aqueous phase at selected invariant points sorted on increasing aqueous [SO_4] (concentration in mol/kg)

No.	Phase assemblages	Ca	Al	Si	CO_3	SO_4	pH
17	Calcite, gibbsite, SiO_2 gel, CSH (I), thaumasite	2.17 E-03	7.16 E-06	5.23 E-03	1.58 E-05	2.49 E-05	10.30
18	Calcite, CSH (I), CSH (II), ettringite, thaumasite	1.97 E-02	9.53 E-07	3.38 E-05	9.96 E-06	3.70 E-05	12.48
19	Calcite, C_2ASH_8, CSH (I), ettringite, thaumasite	4.17 E-03	1.52 E-04	1.04 E-04	1.04 E-05	3.83 E-05	11.84
20	Calcite, C_2ASH_8, gibbsite, CSH (I), thaumasite	2.82 E-03	1.65 E-04	1.70 E-04	1.13 E-05	4.15 E-05	11.66
21	Calcite, portlandite, CSH (II), ettringite, thaumasite	2.20 E-02	5.29 E-07	2.70 E-05	8.93 E-06	4.48 E-05	12.52
22	Calcite, C_2ASH_8, gibbsite, ettringite, thaumasite	3.28 E-03	1.96 E-04	9.13 E-05	1.09 E-05	6.07 E-05	11.73
23	C_2ASH_8, Gibbsite, CSH (I), ettringite, thaumasite	2.86 E-03	1.64 E-04	1.69 E-04	4.50 E-04	1.04 E-06	11.66
24	Gypsum, portlandite, CSH (I), ettringite, thaumasite	3.39 E-02	1.01 E-10	2.59 E-05	3.10 E-08	1.24 E-02	12.49
25	Gypsum, calcite, portlandite, ettringite, thaumasite	3.39 E-02	1.01 E-10	8.95 E-08	8.95 E-06	1.24 E-02	12.49
26	Gypsum, CSH (I), CSH (II), ettringite, thaumasite	3.17 E-02	1.34 E-10	3.22 E-05	2.53 E-08	1.25 E-02	12.44
27	Gypsum, gibbsite, SiO_2 gel, CSH (I), thaumasite	1.60 E-02	4.16 E-06	3.60 E-03	1.03 E-03	1.48 E-02	10.02
28	Gypsum, gibbsite, CSH (I), ettringite, thaumasite	1.58 E-02	6.93 E-06	1.73 E-03	1.05 E-08	1.50 E-02	10.25
29	Gypsum, calcite, gibbsite, SiO_2 gel, thaumasite	1.53 E-02	1.27 E-07	1.46 E-03	8.68 E-05	1.53 E-02	8.51
30	Gypsum, calcite, gibbsite, ettringite, thaumasite	1.54 E-02	7.03 E-06	1.71 E-06	1.08 E-05	1.52 E-02	10.25

Note: C_2ASH_8 is strätlingite. See Ref. 33 for definition of the two hydrogarnets (HG) and two C-S-H formulae included in calculations.

yet reacted chemically with its service environment. Formation of thaumasite now becomes limited by mass considerations, and sulfate, carbonate, and silicate become the most significant components in respect to chemical mass balance limitations on thaumasite formation. However, silicate is less crucial; even assemblages containing free portlandite have sufficiently high silica activities to stabilize thaumasite. Recent decades have seen rising sulfate contents in portland cement as more gypsum (or other sulfate sources) is interground with clinker. However, the amount of sulfate added to modern cements is still insufficient to sustain the relatively high pore fluid sulfate concentrations necessary to stabilize thaumasite. Moreover, in the absence or near absence of carbonate, thaumasite cannot form. Some potential sources of carbonate necessary to form thaumasite may, however, be available. These include:

1. Carbonate in solid solution in the so-called "alkali sulfates," which, as we have shown, are normally alkali sulfate–carbonate solid solutions.[49]

2. Carbonation occurring spontaneously during cement storage.

3. Carbonate contributed from fillers or aggregates.

Of these potential sources, and assuming fresh cement, fillers and aggregates are likely to be quantitatively the major source of carbonate. We do not know if coarse limestone aggregates are sufficiently active in respect to their ability to contribute to thaumasite formation, but the use of limestone fillers, achieved by intergrinding cement clinker and limestone such that the limestone achieves high specific surface area, must be regarded as an area of concern in respect of potential for thaumasite formation: one of the two chemical restrictions necessary to form thaumasite is removed.

However, at low sulfate and carbonate mass fractions, much — perhaps all — of the carbonate and sulfate present will enter the structures of innocuous phases (e.g., sulfate into AFm, carbonate into hemi- or monocarboaluminate). Since we do not at present know the finer details of the fractionation of sulfate and carbonate among the minor phases, we cannot as yet quantify the amounts of carbonate required to initiate thaumasite formation. But most commercial cements, including limestone-filled cements, are probably innocuous in this respect and thaumasite is unlikely to occur spontaneously in sufficient amounts to present problems of dimensional stability in the post-set period.

Concretes in their service environment are often exposed to sulfate and carbonate, and these conditions arguably give rise to most concerns about potential for thaumasite formation. In-service conditions and exchanges are not really considered in this presentation, but clearly the methods described here can be extended, and it can be seen that limestone-filled cements will be most at risk of developing thaumasite if exposed to sulfate-containing groundwaters or other sulfate sources.

Summary and Overview

The author is aware that readers would be turned off by the term "thermodynamics." Consequently, this presentation takes a broad-brush approach. A minimum of equations and formal theory are given; traditionalists may say that not enough has been given. But there seems no point in repeating what is already given in standard textbooks. In any event, software exists to program computers to "know" the relevant equations. Solving even simple thermodynamic problems with pen and paper is tedious and error-prone. So why not let the computer do the hard work, leaving us to get on with the interesting bits? Computers do not, as is sometimes assumed, remove the intellectual challenge of defining the problem, shaping the calculation, taking heed of any kinetically limiting steps and of the results of relevant laboratory or field studies, introducing approximations (if required), and, finally, understanding and applying the results of computations. Application is important; having set out to solve a problem, it is easy to become enmeshed in the mechanics of its solution and in doing so lose sight of the original objective.

Application of modern computational routines enables breakthroughs to be achieved. They enable the thermodynamics of selected aspects of complex multicomponent systems to be calculated. I have often challenged skeptics to say what they wish to know about these complex systems: the usual answer is "everything"! Of course this is impractical; we would be overwhelmed by data output. This is why the application of intellect is still essential to achieve relevance by means of appropriate inputs and to evaluate and place the outputs into context. We have the assurance that the underlying thermodynamic equations are correct. Approximations and assumptions appropriate to cements are, for the most part, user defined, so we can evaluate and control these. Undoubtedly, the marriage of computers and thermodynamics has led to a vast increase in the potential to solve long-standing problems and will revolutionize the way we think about cements and respond to problems concerning their formulation and performance. Direct links between thermodynamic treatments and mechanical properties are still some years into the future, but an important goal for the immediate future is to begin the process of linking thermodynamic treatments to models having as their objective the prediction of engineering properties.

References

1. V. I. Babushkin, G. M. Matveyev, and O. P. Mchedlov-Petrossyan, Thermodynamics of Silicates. Springer-Verlag, Berlin, 1985.
2. C. David Lawrence, "The Constitution and Specification of Portland Cement"; pp. 131–194 in Lea's Chemistry of Cement and Concrete, 4th ed. Edited by P. C. Hewlitt and Ed Arnold. London, 1998.
3. H. A. Bumstead and R. Gibbs van Name, eds., The Scientific Papers of J. Willard Gibbs, Ph.D., LL.D. Dover Publications, London and New York, 1961.
4. Margi Eglinton, "Resistance of Concrete to Destructive Agencies"; pp. 299–342 in Lea's Chemistry of Cement and Concrete, 4th ed. Edited by P. C. Hewlitt and Ed Arnold. London, 1998.
5. "Role of Chemical Binding in Diffusion and Mass Transport"; pp. 129–154 in Ion and Mass Transport in Cement-Based Materials. Edited by R. D. Wooton, M. D. A. Thomas, J. Marchand, J. J. Beaudoin, and J. P. Skalny. Materials Science of Concrete, Special Volume. American Ceramic Society, Westerville, Ohio, 2001.
6. D. Damidot and F. P. Glasser, "Thermodynamic Investigation of the $CaO-Al_2O_3-CaSO_4-CaCl_2-H_2O$ System at 25°C and the Influence of NaCl"; Paper 4-IV-066 in Proc. of the 10th International Congress on the Chemistry of Cements and Concrete (Gothenburg), vol. 4. 1977.
7. S. Stronach and F. P. Glasser, "Modelling the Impact of Abundant Geochemical Components on Phase Stability and Solubility of the $CaO-SiO_2-H_2O$ System at 25°C: Na^+, K^+, and SO_4^{2-}," Adv. Cem. Res., 9, 167–182 (1997).
8. F. P. Glasser, J. Pedersen, K. Goldthorpe, and M. Atkins, "Solubility Reactions of Cement Components with NaCl Solutions: I $Ca(OH)_2$ and C-S-H," Adv. Cem. Res. (in press).
9. Q. Shou, E. E. Lachowski, and F. P. Glasser, "Mettaettringite, a Decomposition Product of Ettringite," Cem. Concr. Res., 34, 703–710 (2004).
10. F .P. Glasser, "Stability of Ettringite"; pp. 43–64 in Internal Sulfate Attack and Delayed Ettringite Formation (Proceedings of the RILEM TC 186 ISA Workshop). Edited by K. Scrivener and J. Skalny. 2004.
11. J. F. Young and W. Hansen, "Volume Relations for C-S-H Based on Hydration Stoichiometries," Mater. Res. Soc. Symp. Proc., 85, 313–322 (1987).
12. C. Dow and F. P. Glasser, "Calcium Carbonate Efflorescence on Portland Cement and Building Materials," Cem. Concr. Res., 33, 147–154 (2003).

13. U. R. Berner, "Modelling the Incongruent Dissolution of Hydrated Cement Minerals," Radiochim. Acta, 44/45, 387–393 (1988).

14. A. Atkinson, J. A. Hearne, and C. F. Knights, "Aqueous Chemistry and Thermodynamic Modelling of CaO-SiO₂-H₂O Gels," J. Chem. Soc., Dalton Trans., pp. 2371–2380 (1989).

15. L. W. Plummer, D. L. Parkhurst, G. W. Fleming, and S. A. Dunkle, "A Computer Programme Incorporating Pitzer's Equations for Calculation of Geochemical Reactions in Brines." Water Resources Investigations Report 88-4153. U.S. Geological Survey, 1988. P. 310.

16. M. Kersten, "Aqueous Solubility Diagrams for Cementitious Waste Stabilization Systems. I. The C-S-H Solid-Solution System," Environ. Sci. Tech., 30, 2286–2293 (1996).

17. D. Sugiyama and T. Fujita, "A Thermodynamic Model of Dissolution and Precipitation of Calcium Silicate Hydrates," Cem. Concr. Res. (in press).

18. G. M. Anderson and D. A. Crear, Thermodynamics in Geochemistry: The Equilibrium Model. Oxford University Press, New York, 1993.

19. D. A. Kulik, V. A. Sinitsyn, and I. L. Karpov, "Predictions of Solid-Aqueous Equilibria in Cementitous Systems Using Gibbs Energy Minimisation: II. Dual Thermodynamic Approach to Estimation of Non-Ideality and End-Member Parameters"; pp. 983–990 in Scientific Basis for Nuclear Waste Management, vol. 506. Edited by I. G. McKinley and C. McCombie. Materials Research Society, Pittsburgh, 1998.

20. V. A. Sinitsyn, D. A. Kulik, M. S. Khodorivsky, and I. K. Karpov, "Prediction of Solid Aqueous Equilibria in Cementitous Systems Using Gibbs Energy Minimisation. I. Mulitiphase Aqueous-Ideal Solid Solution Models"; pp. 953–960 in Scientific Basis for Nuclear Waste Management, vol. 506. Edited by I. G. McKinley and C. McCombie. Materials Research Society, Pittsburgh, 1998.

21. K. Fujii and W. Kondo, "Estimation of Thermochemical Data for Calcium Silicate Hydrate (C-S-H)," J. Am. Ceram. Soc., 66, C220–C221 (1983).

22. E. M. Gartner and H.M. Jennings, "Thermodynamics of Calcium Silicate Hydrate and Their Solutions," J. Am. Ceram. Soc., 70, 743–749 (1987).

23. E. J. Reardon, "Problems and Approaches to the Prediction of the Chemical Composition in Cement/Water Systems," Waste Management, 12, 221–239 (1992).

24. D. A. Kulik and M. Kerston, "Aqueous Solubility Diagrams for Cementitous Waste Stabilization Systems: II, End-Member Stoichiometries of Ideal Calcium Silicate Hydrate Solid Solutions," J. Am. Ceram. Soc., 84, 3017–3026 (2001).

25. S. Börjesson, A. T. Emren, and C. Ekberg, "A Thermodynamic Model for the Calcium Silicate Hydrate Gel Modelled as a Non-Ideal Binary Solid Solution," Cem. Concr. Res., 27, 1649–1660 (1997).

26. C. Plassard, E. Lesnicwska, I. Pockard, and A. Nonat, "Investigation of the Surface Structure and Elastic Properties of Calcium Silicate Hydrate at the Nanoscale," Ultramicroscopy (in press).

27. F. P. Glasser, E. E. Lachowski, and D. E. Macphee, "Compositional Models for Calcium Silicate Hydrate (C-S-H) Gels: Their Solubilities and Free Energies of Formation." J. Am. Ceram. Soc., 70, 1481–1485 (1987).

28. H. M. Jennings, "Aqueous Solubility Relationships for Two Types of Calcium Silicate

Hydrate." J. Am. Ceram. Soc., 69, 614–618 (1986).

29. H. F. W. Taylor, "Nanostructure of C-S-H: Current Status," Adv. Cement-Based Mater., pp. 38–46 (1993).

30. D. Viehland, J. F. Li, L. J. Yuan, and Z. Xu, "Mesostructure of Calcium Silicate Hydrate (C-S-H) Gels in Portland Cement Paste: Short Range Ordering, Nanocrystallinity, and Local Composition Order," J. Am. Ceram. Soc., 79, 1731–1744 (1996).

31. F. P. Glasser and S. Y. Hong, "Thermal Treatment of C-S-H Gel at 1 Bar Pressure up to 200°C," Cem. Concr. Res., 33, 271–279 (2003).

32. D. D. Double and A. Hellawell. "The Hydration of Portland Cement," Nature, 261, 486–488 (1976).

33. M. Atkins, F. P. Glasser, and A. Kindness, "Phase Relations and Solubility Modelling in the CaO-Al$_2$O$_3$-Mg)-SO$_3$-H$_2$O System: For Application to Blended Cements"; pp. 387–394 in Scientific Basis for Nuclear Waste Management XIV. Materials Research Society Proceedings, vol. 212. Materials Research Society, 1991.

34. T. G. Jappy, and F. P. Glasser "Synthesis and Stability of Silica-Substituted Hydrogarnet, Ca$_3$Al$_2$Si$_{3-x}$O$_{12-4x}$(OH)$_{4x}$," Adv. Cem. Res., 4, 1–8 (1991).

35. F. P. Glasser, "Chemical, Mineralogical, and Microstructural Changes Occurring in Hydrated Slag-Cement Blends"; pp. 41–81 in Materials Science of Concrete II. Edited by J. Skalny. American Ceramic Society, Westerville, Ohio, 1991.

36. Q. L. Feng, E. E. Lachowski, and F. P. Glasser, "Densification and Migration of Ions in Blast Furnace Slag–Portland Cement Pastes," Mater. Res. Soc. Proc., 136, 263–272 (1989).

37. E. E. Lachowski, A. Kindness, and F. P. Glasser, "Compositional Development (Solid and Aqueous Phases) in Aged Slag and Fly Ash Blended Cements"; Paper 3-ii, p. 8 in Proc. of the 10th International Congress on the Chemistry of Cements (Gothenburg). 1997.

38. A. Atkins, D. Bennett, A. Dawes, F. P. Glasser, A. Kindness, and D. Read, "A Thermodynamic Model for Blended Cements"; pp. 497–502 in Chemistry of Cements for Nuclear Applications. Edited by P. Barnet and F. P. Glasser. Elsevier, Amsterdam, 1992.

39. K. K. Sagoe-Crentsil and F. P. Glasser, "Steel in Concrete: Part I, A Review of the Electrochemical and Thermodynamic Aspects," Mag. Concr. Res., 41, 205–212 (1989); "Steel in Concrete: Part II, Electron Microscopy Analysis," Mag. Concr. Res., 41, 213–220 (1989).

40. C. L. Page, K. W. J. Treadaway, and P. B. Bamforth, Corrosion of Reinforcement in Concrete. Elsevier Applied Science, 1990.

41. S.-Y. Hong and F. P. Glasser, "Phase Relations in the CaO-SiO$_2$-H$_2$O System to 200°C at Saturated Steam Pressure," Cem. Concr. Res. (in press).

42. C. L. Dickson, D. R. M. Brew, and F. P. Glasser, "Solubilities of CaO-SiO$_2$-H$_2$O Phases at 25°, 55° and 85°C," Adv. Cem. Res., 16, 35–43 (2004).

43. T. Mitsuda and H. F. W. Taylor, "Normal and Anomalous Tobermorites," Mineralogical Mag. (London), 42, 229–235 (1978).

44. Proceedings of the First International Conference on Thaumasite in Cementitious Materials (Garston, Watford, England, June 2002). Edited by I. Holton.

45. K. N. Jallad, M. Santhariam, and M. D. Cohen, "Stability and Reactivity of Thaumasite at Different pH Levels," Cem. Concr. Res., 33, 433–437 (2003).

46. S. Sahu, S. Badger, and N. Thaulow, "Evidence of Thaumatic Formation in Southern California Concrete," Cem. Concr. Compos., 24, 379–384 (2002).

47. N. Gaze and N. J. Crammond, "The Formation of Thaumasite in a Cement: Lime Sand Mortar Exposed to Cold Magnesium and Potassium Sulfate Solutions," Cem. Concr. Res., 30, 209–222 (2000).

48. D. Damidot, S. J. Barnett, F. P. Glasser, and D. E. Macphee, "Investigation of the CaO-Al_2O_3-SiO_2-$CaSO_4$-$CaCO_3$-H_2O System at 25°C by Thermodynamic Calculations," Adv. Cem. Res., 16, 69–76 (2004).

49. T. I. Barry and F. P. Glasser "Calculation of Portland Cement Clinkering Reactions," Adv. Cem. Res., 12, 19–28 (2000).

Tricalcium Silicate Hydration: A Historical Overview

J. Francis Young

Understanding the hydration of tricalcium silicate began with the first studies by Le Chatelier. Much of the foundation for this was laid in his doctoral thesis published in 1887, the first seminal publication on the subject. While Le Chatelier had determined its composition, more extensive studies of hydration required a convenient synthesis, which was not attained until 1917. By 1935 researchers had established the basic hydration stoichiometry, although the overall composition of C-S-H was still uncertain. The work of Brunauer and Kantro confirmed detailed stoichiometry and demonstrated its variability with different conditions of hydration. Electron microscopy and microanalysis were required to determine the physical relationships of the hydration products (microstructure) and the inherent variability of C-S-H within the microstructure. Modern analytical techniques have allowed mechanistic descriptions of the hydration process to be developed. A summary of the current knowledge of the hydration of tricalcium silicate is provided.

Introduction

A historical review of tricalcium silicate hydration is most appropriate for a number of reasons. First, C_3S* is the major constituent of portland cement, and controls setting and early strength development. Second, the hydration of C_3S has been a major research interest of many of our prominent scientists since Le Chatelier first identified and studied this phase. Such luminaries as Bates, Bogue, Lerch, Flint, Thorvaldsen, Brunauer, and Copeland have studied the hydration of C_3S. Third, it is important for young researchers to develop an appreciation of the history of their field and how our present day knowledge has evolved.

C_3S has been recognized as the principal active ingredient of cement since Le Chatelier[1] showed that free lime was absent in a good quality portland cement, the presence of Ca aluminates was not definite, and C_2S was considered to be nonhydraulic. In 1917 Bates and Klein[2] confirmed what Le Chatelier had implied: that C_3S has all the essential properties of port-

---

This paper is based on the Della M. Roy lecture presented at the Cements Division meeting of the American Ceramic Society in 2003.

*Conventional shorthand notation is used in this paper: C = CaO; S = SiO_2; H = H_2O, etc.

Table I. Progress in the study of the hydration of tricalcium silicate

Early period: Pre-1950
Progress is slow owing to limited experimental methods.
Optical microscopy and bulk chemical analysis are the principal methods.
Middle period: 1950–1990
More rapid progress is made, owing to a wider range of analytical methods:
X-ray diffraction, electron microscopy and microanaylsis, solid-state NMR spectroscopy.
Late period: Post-1990
More sophisticated development of the above analytical techniques.
Computer simulation modeling is introduced.

Table II. Early synthesis of tricalcium silicate

Le Chatelier (1887)
$2C + S + CaCl_2 \rightarrow C_2S \cdot CaCl_2$
$C_2S \cdot CaCl_2 + H_2O \rightarrow C_3S + HCl$
Newberry (1903)
$3C + S \xrightarrow{\text{fusion}} C_3S$
Bates and Klein (1917)
Repeated sintering of $(3C + S)$ with intermediate grinding

land cement, with initial and final set and no unsoundness. Its study as a model compound has helped us to understand the more complex system.

The study of C_3S has been divided into three arbitrary periods (Table I) in the same way that C_3S hydration can be divided into three periods. The early and middle periods will primarily be discussed since the late period covers the more recent work. A summary of the current views on C_3S hydration concludes this chapter.

Formation and Composition of C_3S

Early nineteenth century workers had concluded that portland cement consisted of hydraulic calcium aluminates and perhaps aluminosiliciates. Calcium silicates appeared to be nonhydraulic since only γ-C_2S could be synthesized. In 1856 Rivot[3] suggested that cement consisted of C_3S and CA, and this view was widely adopted. However, Le Chatelier[1] observed:

This theory is the one . . . generally taught; it is, however, in reality, only a hypothesis which owes its life to the name of its author much more than the experimental truths on which it is based.

During his thesis work Le Chatelier became convinced of the existence of C_3S, but had trouble synthesizing it by direct calcination of lime-silica mixes. Eventually he successfully used a chloride flux approach (Table II), but the decomposition of the calcium chlorosilicate was slow and incomplete. Later Newberry* formed C_3S by complete fusion of lime-silica mixtures. Bates and Klein[2] in 1917 were probably the first to introduce the repeated sintering of stoichiometric lime-silica mixes with intermediate grinding to produce pure C_3S. This approach was gradually refined[4,5] and had become the standard method by 1935[6]; it remains so to this day.

Le Chatelier identified C_3S by the analysis of hard grains ("grappiers"), which under the optical microscope were seen to be low in interstitial material and did not dust (i.e., they were low in C_2S). Working independently, without knowledge of Le Chatelier's work, Tornebohm[7] used density fractionation to isolate alite and analyze it; he proposed the formula $9(C_3S)\cdot C_9A_2$. (It was Tornebohm who coined the terms "alite" and "belite.") While some supported Le Chatelier's claim of C_3S as a distinct compound, others considered it to be a solid solution, notably Michaelis (see Ref. 8). In 1904–1905 Richardson proposed $6(C_3S)\cdot C_3A$ as the formula for alite. It was not until 1952 that Jeffery[9] identified the polymorphic transformations for pure C_3S and proposed that alite was stabilized, monoclinic C_3S with the approximate formula $C_{54}S_{16}AM$, which has a much lower aluminum content than previous formulations. Comparisons of analyses by the various investigators are summarized in Table III. Le Chatelier's analyses are reasonably accurate, and Jeffery's formulation is a good approximation to modern analytical data. After 1965 there were several independent applications of electron microanalysis to analyze individual crystals of alite.[10,11] Average values from all of these studies were calculated by Taylor.[11]

Stoichiometry of C_3S Hydration

Progress in determining the stoichiometry of C_3S is summarized in Table IV. The equation proposed by Le Chatelier was analogous to the hydration of B_2S, which he also studied. He recognized the problem of the variable composition for C-S-H, which he believed arose from adsorption of lime.

---

*As quoted in Le Chatelier; which Newberry he is referring to, W. B. or S. B., is not clear. However, C_3S cannot be crystallized from a melt of its own composition.

Table III. Analysis of tricalcium silicate in portland cement (alite)

Investigator	SiO_2	CaO	Al_2O_3	Fe_2O_3	MgO	C/S
LeChatelier	26.0	66.0	3.0	1.2		2.75
	26.0	66.0	3.5	0.8		2.75
	24.0	69.0	2.7	0.3		3.08
	25.5	68.0	3.6	0.7		2.85
Tornebohm	19.5	67.6	7.8		3.0	3.71
Richardson	30.6	71.8	8.6			2.50
Jeffery	23.3	73.3	2.4			3.37
Taylor	25.2	71.6	1.0	0.7	1.1	3.04

Table IV. Stoichiometry of C_3S hydration

Le Chatelier (1887)
$C_3S + 4.5H \rightarrow CSH_{2.5} + 2CH$
Newberry and Smith (1903)
$C_3S + yH \rightarrow C_xSH_z + (3-x)CH$
Lea and Desch (1935)
As above with $x = 1.5$–2.0
Brunauer and Kantro (1966)
$C_3S + 2.5H \rightarrow C_{(1.5+m)}SH_{(1+m)} + (1.5-m)CH$
Taylor (1990; average of many studies)
$C_3S + (2.6$–$2.8)H \rightarrow C_{1.7}SH_{1.3-1.5} + 1.3CH$
Young and Hansen (1987; saturated conditions)
$C_3S + 5.3H \rightarrow C_{1.7}SH_{4.0} + 1.3CH$

By reacting colloidal silica with saturated lime water he was able to form $C_{1.3}SH_x$ initially, which aged with time to $C_{1.7}SH_x$. Lime was readily extracted by water to form $CSH_{2.5}$. Later work by his contemporaries suggested that the C/S value of C-S-H was greater than 1.0 and might be as high as 2.0. In 1935 Lea and Desch[6] summed up the contemporary view. CH crystallized soon after mixing with water, and C-S-H formed as gelatinous precipitate around the original grains with incomplete hydration. The value of x appeared not to be fixed. By 1947 Bogue[8] considered that the existence of $C_3S_2H_x$ was a strong possibility, although the evidence for it was still inconclusive.

The definitive research by Brunauaer, Kantro, and coworkers (summarized in Ref. 12) in the late 1950s and early 1960s put the stoichiometry of hydration on a solid footing. They first suggested that C_3S_2H was the formula for C-S-H, but later modified this to allow for variability in the values of C/S up to 1.7. Composition depended on the time and temperature of hydration and w/c ratio. During the early stages of hydration the range of composition is highly variable with C/S = 1.0–3.0; at later times it narrowed to 1.4–2.0. These studies depended on bulk analysis of crystalline compounds with the composition of C-S-H determined from mass balances. There was considerable debate about the merits of determining CH contents from QXRDA, thermal analysis, or organic extraction, since there are problems with each method giving slightly different results.

Nevertheless, the variability found from bulk analysis has been confirmed by microanalytical techniques with sampling down to cubic nanometer volumes. The value of the C/S ratio is now accepted to be an average value of 1.7 regardless of the water content of the preparation. In the late 1980s, Taylor,[11] working with cement, and Young and Hansen,[13] working with C_3S, addressed this issue in terms of stoichiometries. Their joint conclusion was that the best representation of C_3S hydration is the last equation, which indicates a much higher water content (and a lower density) when formed under staturated, or near saturated, conditions.

Brunauer and Kantro[12] were the first to consider the water content of C-S-H systematically. They recognized that because C-S-H has no definite hydrates it loses water continuously on dehydration. Since the water content depended on the water vapor pressure during drying, they settled on D-drying as providing samples that are fully dried. Under these conditions H/S = C/S – 0.5. Until recently all formulas for C-S-H were reported for the D-dried condition as the reference state, even though this is not the condition in concrete. Such formulae are in gross error if one tries to calculate volumes in near-saturated pastes for concrete in order to calculate porosities. Saturated C-S-H will have a much higher water content and a lower density. This had been implicitly recognized by Powers and Brownyard[14] in 1947 and was pointed out by NRC scientists in Canada in the early 1970s.[15] The composition proposed by Young and Hansen for saturated pastes gives volume calculations that are in good agreement with those made by Powers[14] and Taylor[11] for portland cement.

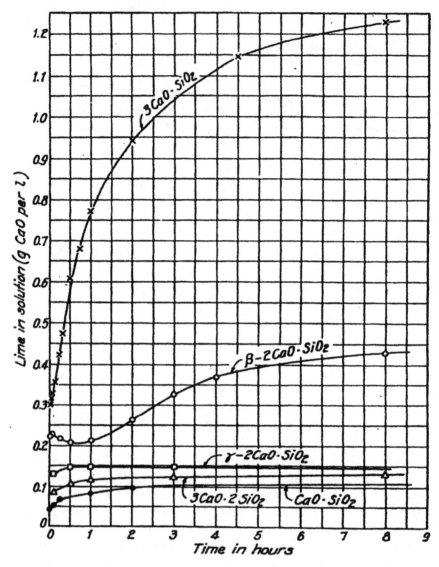

Figure I. Rate of dissolution of calcium silicates.[18]

Solution Composition

The focus on the composition of the solid phases in the 1950s and 1960s directed attention from role of the solution composition during hydration. Attention to this problem began again in the 1970s and 1980s, but in fact it was studied much earlier. The rapid release of lime into solution with a

Table V. Effect of pH on the dissolution of C_3S (w/s = 0.50)[16]

Solution concentration	pH	pH (C_3S)
Water	6.72	12.44
0.21 g CH/L	11.83	12.43
0.568 g CH/L	12.18	12.45
1.152 g CH/L (saturated)	12.37	12.45
NaOH solutions	11.93	12.58
	12.20	12.61
	12.58	12.67
	12.94	13.03
	13.72	13.72

concomitant rise in pH was noted by Lerch and Bogue,[16] Thorvaldson and Vignusson,[17] and Flint and Wells[18] (see Fig. 1). Lerch and Bogue found that the solution became supersaturated with respect to lime within the first hour (and in portland cement within 8 h), and that this happened even when dissolved in lime solutions. C_3S dissolved even in alkaline solutions until a pH of 13.7 was reached (see Table V). This early work showed that the composition of the product precipitated from solution depends on the lime concentration reached. At low lime contents the composition is close to C/S = 1.0, and when approaching saturation it is close to 2.0. Grutzeck,[19] in his pH-controlled experiments 60 years later, found similar results.

Forsen[20] and Hedin[21] studied dilute suspensions of C_3S in water (0.24 g/L) and observed congruent dissolution. Forty years later there was a lively debate[22] regarding whether C_3S dissolves congruently (the Dijon school) or incongruently (Skalny and Young). It is now generally agreed that congruent dissolution probably occurs even in pastes, but that precipitation of early products and modification of the C_3S surface occurs extremely rapidly, so that the rate of dissolution is changed within less than a minute. The overall effect is effective incongruent dissolution under paste conditions. In stirred dilute suspensions (w/s = 5000; 0.2 g C_3S/L) Damidot et al.[23] dissolved C_3S within 30 min, from which Gartner[24] calculated a dissolution rate of $>10^{-3}$ mol/s. At w/s = 20 it was estimated that complete dissolution would require several days owing to the blocking effects of surface precipitates.

Nonat and coworkers[25,26] at Dijon have extensively studied the hydration of dilute, stirred suspensions under conditions in which CH did not precipitate. The use of increasingly high concentrations of lime on solution slows hydration initially, but subsequent hydration is greater. Their data are consistent with C-S-H growing from a relatively small number of nuclei as laminar morphologies parallel to the surface. The initial kinetics are consistent with those determined from calorimetry of low w/c pastes. Studies of solution compositions at w/s ratios more comparable to pastes did not occur until the 1970s. Young and coworkers[27] showed that the concentration of CH in solution rises to a value nearly twice that for saturation. The value of CH_{max} is independent of w/c below about 50.[28] (See Fig. 2.)

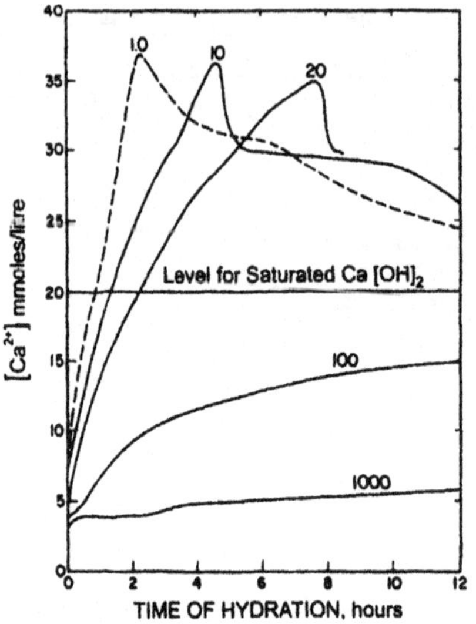

Figure 2. Dissolution of tricalcium silicates at different w/s. (Reproduced with permission from Ref. 45.)

Microstructure Development

Setting and Hardening

It was implied by Le Chatelier that the setting and hardening of portland cement is due to the hydration of C_3S. Bates and Klein[2] and later Bogue and Lerch[29] confirmed that normal setting and hardening are controlled by C_3S. C_3A may react to cause flash set if not properly controlled by gypsum. This view is still held today, although it was revisited by Locher and coworkers[30] in the late 1970s. The nature of the setting process was the first great debate of cement chemistry and is extensively reviewed in Ref. 8. On the basis of his work with plaster, Le Chatelier proposed that setting and hardening were due to the formation of small elongated crystals of C-S-H, which provide a high degree of local cohesion and strength. Michaelis, on the other hand, advanced the theory that setting and hardening are caused by the dehydration of the initial hydrogel, which is dehydrated as water is consumed by further hydration. Crystallization processes (e.g., CH) contribute only to long-term hardening. These two theories were the subject of intense debate for about 25 years, until the first international symposium in 1919. In the end a compromise was reached that there was an element of truth in both points of view. Desch[31] summed it up as follows:

I think it is very clearly established . . . that we are dealing largely with a difference in terms; that whether one regards the jelly as a mass of extremely minute interlacing particles or not is not of very much importance. At any rate, the essential point is that in the colloidal substance the particles are extremely small, and therefore the surface forces are very important. When you come to crystals of perceptible size, such as the crystallization of sodium sulfate, the surface forces are very small relative to the forces of cohesion. In the case of the ultramicroscopic particles in the colloid, the surface forces are large in proportion to the forces of cohesion, and when that fact is fairly grasped it is seen that there is no great question of fact at issue between the two views.

Baykoff,[32] in a paper presented to the French Academy of Science in 1926 and read by Le Chatelier, proposed that hardening takes place in three stages:

The first is that of solution, during which the liquid is saturated progressively with the different soluble elements.

The second is the "colloidation" during which all the products of the chemical reaction form in the colloidal state. That corresponds to the beginning of set.

The third is that of crystallization, during which the gels are transformed into crystalline aggregates. This is the period of hardening, properly speaking.

The study of setting and hardening was hampered by the lack of modern characterization techniques that would allow a good appreciation of how the hydration products combined together to form the hardened mass — the microstructure. Optical microscopy was the principal tool, but could be misleading even in the hands of experienced microscopists. Le Chatelier thought that C-S-H was crystalline, although later investigators concluded that it was amorphous (confirmed eventually by XRD) and formed a layer around the C_3S particle. The invention of the conduction calorimeter by Carlson[33] provided a useful tool that Lerch[34] used to understand the role of sulfate in the setting of cement. Later, Young et al.[27] were able to correlate the onset of the major heat evolution peak with phenomena associated with setting.

Table VI. C-S-H formed in the microstructure (after setting)

Outer product C-S-H
Formed mostly outside the original boundaries of the solid particles.
Contributes to setting and early hardening.
Inner product C-S-H
Formed mostly within the original particle volume by slow, diffusion-controlled mass transfer

Microstructural Studies

The development of SEM was necessary to resolve the details of the microstructure. The new technique appears to have first been applied to C_3S by Sierra[35] in 1968. Subsequently, systematic studies on C_3S paste microstructure[36,37] in the early 1970s helped to build up a qualitative picture of microstructural development and its relationship to properties. These could be compared with contemporary studies on portland cement,[38] confirming the central role of C_3S, but also showing important differences between the two. For example, hollow shell hydration does not occur with pure C_3S or alite.

The development of TEM was also important in probing the microstructural aspects down to the nanoscale. Morphology and composition of C-S-H within the microstructure can be determined precisely. The broad conclusions determined from studies of the bulk phases have been shown to be largely correct, but with finer details now revealed. An excellent overview of the electron-optical studies of hydrated pastes of cement and tricalcium silicate was published recently by Richardson.[39] Improvements in instrument capabilities and improved microanalytical capabilities have allowed quantitative descriptions of the microstructure. Finally, the huge advances in computing have allowed computer-based modeling based on reaction stochiometries to create quite realistic digital models of the microstructure, which can be used to develop refined quantitative relationships between microstructure and properties.

C-S-H can be divided into two types (Table VI). One type formed early and mostly outside the original boundaries of the solid particles contributes to setting and early hardening. The other type formed later, mostly within the original particle volume, as a result of slow, diffusion-controlled mass transfer, contributes to later hardening. These distinctions are based on electron microscopy.[40] Two types of C-S-H have been distinguished also by Jennings and coworkers[41,42] using density and surface area. Whether these two classifications are identical remains to be seen.

Table VII. First proposed mechanism of C₃S hydration

Stage	Action
I	Hydrated "skin" formed on the surface of C_3S with a C/S ratio approaching 3.0
II	Conversion to a second intermediate with a low C/S ratio, and liberation of lime
III	Stable form of C-S-H attained by uptake of lime; forms a coating around C_3S particles

Mechanisms of Hydration

Early Studies

Little attention was paid to the mechanisms by which C_3S hydrated during the early period. However, Le Chatelier had recognized that the onset of heat evolution was delayed and signaled the onset of setting. Brunauer and Kantro[12] proposed a three-stage process (Table VII). The first step forms a hydrated "skin" on the surface of C_3S with a C/S = 3.0. This converts to a second intermediate with a low C/S and the liberation of lime. Stein and coworkers[43] have proposed a similar process. Subsequent uptake of lime again converts it to the stable form of C-S-H around the C_3S particles. Diffusion control takes over when the coating becomes sufficiently thick during the third stage. Diffusion control later in hydration was already recognized and had been mathematically modeled by Taplin in 1960.[44] He was the first to introduce the concept of two types of C-S-H (inner and outer prod· ts) with different diffusion coefficients to explain kinetic data, although he had no direct evidence at the time for such a distinction. ESCA studies about 20 years later, which are summarized in Ref. 45, determined that the C/S ratio of the initial product was less than 3.0. The absence of definite hydrate morphologies led to the suggestion that the initial dissolution was incongruent, releasing lime and forming a silica-rich surface. Under paste conditions there is evidence for a very rapid precipitation soon after contact with water followed by a gradual buildup of intermediate product.

Multistage Hydration

The current concept of five stages of hydration in cement hydration was first proposed by Kondo and coworkers[46] in 1969, and it also applies to hydrating C_3S. The concept of the dormant period (Stage 2), now called the induction period, was introduced. It is clear that Stage 1 is part of an initial peak. The concept of early, middle, and late hydration (Fig. 3) was introduced by Skalny, Young, and Jawed[45] to emphasize that diffusion limited early hydration rates in a similar manner to later hydration. Time scales are

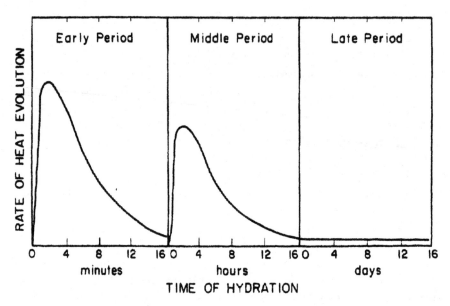

Figure 3. Different periods of tricalcium silicate hydration. (Reproduced with permission from Ref. 45.)

minutes for the early period (encompassing Stages 1 and 2) and hours for the middle period (encompassing Stages 3 and 4). The late period is the same as Stage 5, the diffusion-controlled "steady state" period of hydration. This viewpoint also emphasizes the presence of the initial peak. The formation of C-S-H within these periods is summarized in Table VIII, although the distinction between the middle and later periods is probably not as clear cut as this classification might imply.

So, Stage 0 has now been added (Fig. 4) as the initial dissolution of C₃S before it is limited by surface modification and precipitates. Type E C-S-H is an intermediate form, observed in wet-cell TEM during Stage 1, which later converts to outer product C-S-H. Jennings's analysis[47,48] of published solubility data gave a thermodynamic underpinning to this intermediate product being a distinct phase. Figure 5 shows that "metastable" C-S-H(m), which forms during Stages 1 and 3, is indeed less stable than the outer product C-S-H(s), which forms after the end of the induction period. Barret and Bertrandie[49] have used similar distinctions, although they employed a different nomenclature. C-S-H(m) is presumably the "hydrated skin" proposed by various investigators[12,43] and there are structural differences between C-S-H(m) and C-S-H(s).

Table VIII. C-S-H formed during the hydration of C₃S

	Designation	References
Early period		
Intermediate C-S-H	"Hydrated skin"	12, 43
	Type E C-S-H	see 24
	C-S-H(m)	48
Middle period		
Outer product C-S-H	Second intermediate	12
	C-S-H(s)	48
	Low-density C-S-H	41, 42
	Op C-S-H	40
Late period		
Inner product C-S-H	Stable C-S-H	16
	High-density C-S-H	41, 42
	Ip C-S-H	40

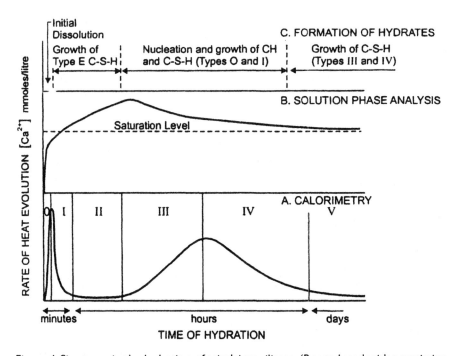

Figure 4. Six stages in the hydration of tricalcium silicate. (Reproduced with permission from Ref. 24.)

Termination of the Induction Period

Another vigorous debate has been over the phenomenon that triggers the end of the induction period. Is the determining event nucleation of crystalline CH, or the conversion of early C-S-H to its stable form? Gartner and Gaidis[50] have noted that Stage 2 is highly idealized, and in practice this stage effectively represents a minimum between Stages 1 and 3, except in very retarded systems. Their reexamination of the kinetics during Stage 3 suggests an exponential growth process, which is not inconsistent with the structure and morphology of C-S-H, and was earlier proposed for the hydration of dilute suspensions.

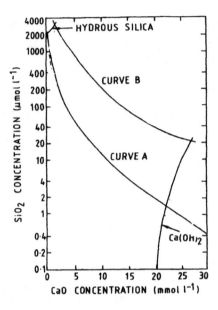

Figure 5. Phase diagram for hydrating tricalcium silicate.[47]

Nevertheless, the need for CH to form in paste hydration is undeniable, and the difficulty of this nucleation step is emphasized by the high degree of supersaturation attained before nucleation occurs. Furthermore, how can one poorly ordered phase nucleate from another? An approach that removes this apparent dichotomy is based on short-range order concepts. "Stable" C-S-H is considered to have a defect tobermorite/jennite structure (depending on composition) with a distorted CH structure as the central layer. Thus both CH and C-S-H form from a common nucleus: that of a CH layer with adsorbed disilicate ions. Most of these nuclei grow into C-S-H structures with only a relative few into CH crystals, the relative amounts depending on the degree of supersaturation, as well as entropy considerations.

Based on the above discussion, the current view of the hydration of C_3S can be summarized as shown in Table IX.

Concluding Remarks

Le Chatelier first put hydration chemistry on a rational footing 125 years ago. It is amazing how accurate his observations were in light of the crude

Table IX. Summary of hydration of tricalcium silicate

$$C_3S + 5.3H \rightarrow C_{1.7}SH_{4.0} + 1.3CH$$

Stage	Processes operating	Rate of reaction
0	Initial congruent dissolution	Extremely rapid (seconds)
1	Rapid dissolution blocked by surface chemistry modification	Rapid (minutes)
2	Induction period; formation of intermediate (Type E) C-S-H	Moderate (1–2 hours)
3	Onset of setting and hardening as CH and C-S-H nucleate in an autocatalytic process	Moderate (hours)
4	Onset of diffusion-controlled kinetics as C-S-H coating thickens	Slow (hours–days)
5	"Steady state"; slow, continued hydration through diffusion-controlled mass transfer	Very slow (weeks–years)

tools he had at his disposal. Subsequent progress has been uneven, but steady. By 1935, when the first text on cement chemistry was published,[6] the broad outline of C_3S hydration was already established, although the stoichiometry of hydration had not been pinned down. And little could be said about the development of the paste microstructure or variability of the C-S-H composition before the advent of electron microscopy. Nevertheless it is salutary to see how much progress could be made in unraveling the chemistry without the use of modern analytical tools, which we now take for granted and which make the task so much easier.

During the middle period a detailed knowledge of the stoichiometry of C_3S hydration and how it develops the paste microstucture was established. We also obtained a quantitative handle on the kinetics and laid the foundations for a reasonable mechanistic understanding of how hydration proceeds. In recent years (the late period) we have refined these ideas and concepts. While there is still room for further argument and debate on many points, the big issues have essentially been resolved.

Acknowledgments

I would like to acknowledge some of my colleagues who at various times have worked with me on C_3S hydration or who have debated at length the issues I have talked about: R. L. Berger, F. V. Lawrence, L. E. Copeland, S. Diamond, J. P. Skalny, H. F. W. Taylor, W. Hansen, and E. M. Gartner.

References

1. H. Le Chatelier, *Experimental Researches on the Constitution of Hydraulic Mortars.* Translated by J. L. Mack. McGraw, New York, 1905.
2. P. H. Bates and A. A. Klein, Technical Paper 78. National Bureau of Standards, 1917.
3. Rivot, *Ann. des Mines 5e serie,* **IX**, 505 (1856).
4. W. C. Hansen, *J. Am. Ceram. Soc.,* **11**, 68 (1928).
5. J. Weyer, *Zement,* **20**, 692 (1931).
6. F. M. Lea and C. H. Desch, *The Chemistry of Cement and Concrete,* 1st ed. Arnold, London, 1935. P. 49.
7. A. E. Tornebohm, "Uber die Petrographie des Portland Zements," *Tonind. Ztg.,* **21**, 1148 (1897).
8. R. H. Bogue, *The Chemistry of Portland Cement.* Reinhold, New York, 1947. Pp. 44–60.
9. J. W. Jeffery, "Tricalcium silicate"; pp. 30ff. in *Proc. 3rd Intern. Symp. Chemistry of Cement* (London, 1952).
10. C. D. Lawrence, "The Constitution and Specifications of Portland Cement"; pp. 157–161 in Lea's Chemistry of Cement and Concrete, 4th ed. Edited by P. C. Hewlett. Arnold, London, 1998.
11. H. F. W. Taylor, *Cement Chemistry,* 2nd ed. Thomas Telford, London, 1997.
12. S. Brunauer and D. L. Kantro, "The Hydration of Tricalcium Silicate and Dicalcium Silicate from 5°C to 50°C"; pp. 287–309 in *The Chemistry of Cements.* Edited by H. F. W. Taylor. Academic, London, 1964.
13. J. F. Young and W. Hansen, "Volume Relationships for C-S-H Formation Based on Hydration Stoichiometries," *Mater. Res. Soc. Symp. Proc.,* **85**, 313–322 (1987).
14. T. C. Powers and T. L. Brownyard, "Physical Properties of Cement Pastes"; p. 597 in *Proceedings of the 4th International Symposium on Chemistry of Cement,* vol. II. NBS Monograph No. 42. Washington, 1962.
15. R. F. Feldman and V. S. Ramachandran, "A Study of the State of Water and Stoichiometry of Bottle Hyrdated Ca_3SiO_5," *Cem. Concr. Res.,* **4**, 155–166 (1974).
16. W. Lerch and R. H. Bogue, *J. Phys. Chem.,* **31**, 1627 (1931). Quoted in Ref. 8, pp. 382–384.
17. T. Thorvaldson and V. A. Vigfusson, *Trans. Royal Soc. Canada,* **22**, 423 (1934). Quoted in Ref. 8, p. 384.
18. E. P. Flint and L. S. Wells, *J. Res. Natl. Bur. Stds.,* **12**, 751 (1934). Quoted in Ref. 8, pp. 385–388.
19. M. Grutzeck, A. Benesi, and B. Fanning, "^{29}Si Magic Angle Spinning Nuclear Magnetic Resonance Study of Calcium Silicate Hydrate," *J. Am. Ceram. Soc.,* **72**, 665–668 (1989).
20. L. Forsen, *Proceedings of the 2nd Symposium on Chemistry of Cement.* Stockholm, 1938. P. 298. Quoted in Ref. 8, pp. 389–390.
21. R. Hedin, *Proc. Swedish Cem. Concr. Inst.,* **3** (1945). Quoted in Ref. 8, pp. 389–390.
22. J. Skalny and J. F. Young, "Mechanisms of Portland Cement Hydration"; pp. II-1/3–II-145 in *Proceedings of the 7th International Congress on Chemistry of Cement,* vol. 1. Paris, 1980.
23. D. Damidot, A. Nonat, and P. Barret, *J. Am. Ceram. Soc.,* **73**, 3319 (1990).
24. E. M. Gartner, J. F. Young, D. Damidot, and I. Jawed, "Hydration of Portland Cement"; pp. 57–113 in *Structure and Performance of Cements,* 2nd ed. Edited by J. Bensted and

P. Barnes. Spon Press, London, 2002.

25. D. Damidot and A. Nonat, *Adv. Cem. Res.,* **6**, 27 (1994).

26. S. Gauffinet, E. Finot, and A. Nonat, *Proceedings of the 2nd International RILEM Workshop on Hydration and Setting.* Dijon, 1997.

27. J. F. Young, H.-S. Tong, and R. L. Berger, "Compositions of Solutions in Contact with Hydrating Tricalcium Silicate Pastes," *J. Am. Ceram. Soc.,* **60**, 193–198 (1977).

28. D. Menetrier, D.Sc. Thesis, University of Dijon, 1977. (See Ref. 45.)

29. R. H. Bogue and W. Lerch, *Ind. Eng. Chem.,* **26**, 837 (1934).

30. F. W. Locher, W. Richartz, and S. Sprung, *Zement-Kalk-Gips,* **29**, 435 (1976).

31. C. H. Desch, *Trans. Faraday Soc.,* **14** [1] 67 (1919). (Proceedings of the 1st Symposium on Chemistry of Cement, London.) Quoted in Ref. 8, pp. 364–365.

32. M. Baykoff, *Compte Rendu,* **63**, 1324 (1926). Quoted in Ref. 8, p. 365.

33. R. W. Carlson, *Proc. Amer. Soc. Test. Mater.,* **34**, 322 (1934).

34. W. Lerch, *Proc. Amer. Soc. Test. Mater.,* **46**, 1252 (1946).

35. R. Sierra, "Etude au microscope electronique de l'hydratation des silicates calciques du ciment Portland," *J. Microscopie,* **7**, 491–508 (1968).

36. M. Collepardi and B. Marchese, "Morphology and Surface Properties of Hydrated Tricalcium Silicate Pastes," *Cem. Concr. Res.,* **2**, 57–65 (1972).

37. F. V. Lawrence Jr. and J. F. Young, "Studies on the Hydration of Tricalcium Silicate. I. Scanning Elecron Microscopic Examination of Microstructural Features," *Cem. Concr. Res.,* **3**, 149–161 (1973).

38. S. Diamond, "Cement Paste Microstructure — An Overview at Several Levels"; pp. 2–30 in *Proceedings of the Conference on Hydraulic Cement Pastes: Their Structure and Performance.* Sheffield, UK, 1976.

39. I. G. Richardson, "Electron Microscopy of Cements"; pp. 500–556 in *Structure and Performance of Cements,* 2nd ed. Edited by J. Bensted and P. Barnes. Spon Press, London. 2002.

40. I. G. Richardson and G. W. Groves, "Microstructure and Microanalysis of Hardened Ordinary Cement Pastes," *J. Mater. Sci.,* **28**, 265–277 (1993).

41. H. M. Jennings, "A Model for the Microstructure of Calcium Silicate Hydrate in Cement Paste," *Cem. Concr. Res.,* **30**, 102–115 (2000).

42. P. D. Tennis and H. M. Jennings, "A Model for Two Types of Calcium Silicate Hydrate in the Microstructure of Portland Cement Pastes," *Cem. Concr. Res.,* **30**, 855–863 (2000).

43. J. G. M. de Jong, H. N. Stein, and J. M. Stevels, *J. Appl. Chem (London),* **17**, 246 (1967).

44. J. H. Taplin , Proceedings of the 5th International Symposium on Chemistry of Cement, Tokyo, 1968. Vol. 2, pp. 249, 337.

45. I. Jawed, J. Skalny, and J. F. Young, "Hydration of Portland Cement"; pp. 237–318 in *Structure and Performance of Cements,* 1st ed. Edited by P. Barnes. Applied Science, London, 1983.

46. R. Kondo and M. Daimon, *J. Am. Ceram. Soc.,* **52**, 503 (1969).

47. H. M. Jennings, *J. Am. Ceram. Soc.,* **69**, 614 (1986).

48. E. M. Gartner and H. M. Jennings, *J. Am. Ceram. Soc.,* **70**, 743 (1987).

49. P. Barret and D. Bertrandie, *J. Am. Ceram. Soc.,* **71**, C113 (1988).

50. E. M. Gartner and J. M. Gaidis, "Hydration Mechanisms I"; pp. 95–126 in *Materials Science of Concrete I.* Edited by J. Skalny. American Ceramic Society, Westerville, Ohio, 1989.

Advanced Methods for Investigating Cement Hydration

R. James Kirkpatrick

This chapter discusses the current and potential contributions of a variety of rapidly developing and less commonly used experimental methods used to investigate the hydration of portland cement and the structure and chemical properties of the hydration products. Emphasis is placed on neutron and X-ray scattering and spectroscopic techniques that require large-scale facilities such as neutron sources and third-generation synchrotrons, because application of these techniques in cement and concrete science is advancing rapidly. Other experimental techniques, such as Raman spectroscopy, X-ray photoelectron spectroscopy, electron energy loss spectroscopy, and acoustic microscopy also have important capabilities.

Introduction

Understanding of the chemical processes involved in the hydration of portland cement and the molecular and nanoscale structure and chemical properties of the hydration products advanced greatly during the latter half of the twentieth century through application of a wide variety of experimental techniques. These included scanning and transmission electron microscopy, X-ray diffraction, calorimetry, electron probe chemical analysis, NMR spectroscopy, thermal analysis, and various probes of pore size distribution. Portland cement is an extraordinarily complex system, however, and despite these efforts much remains to be learned before it can be controlled and manipulated with the subtlety of metals and high-temperature ceramics. Key unresolved issues include the molecular-scale mechanisms by which water attacks unhydrated cement grains; the mechanisms by which early hydration products form; the detailed molecular- and nanoscale structures present in C-S-H; the size distributions and temporal evolution of these structures; the amount, distribution, temporal evolution, and behavior of solid surface and pore volume; and the distribution, chemical environments, and chemical behavior of minor species including SO_4^{2-} and a wide range of species related to nuclear and hazardous waste sequestration.

In the past decade there has been increasing application of advanced experimental techniques to understanding cement hydration and the hydration products, and this paper discusses current and potential contributions of some of these methods. The most notable of these are X-ray and neutron scattering and spectroscopic techniques that take advantage of the increasing availability of high-intensity X-rays from third-generation synchrotrons and increased flux from advanced neutron sources. Other laboratory-based techniques, such as Raman spectroscopy and X-ray photoelectron spectroscopy (XPS), also hold significant potential. For many of these techniques, reference is made here to key review papers and important recent research papers, but comprehensive review of all contributions using the techniques is beyond the scope of this paper.

Synchrotron-Based Methods

Synchrotrons facilities are now available worldwide and are making transformative contributions to fields as diverse as materials science, biochemistry, and geoscience. There has been significant recent use of some synchrotron-based techniques in cement and concrete science, but many approaches have not been explored and the future potential is great. The various papers in Fenter et al.[1] provide excellent reviews of the operation of synchrotrons and many of the experiments that can be done with them. The X-ray scattering and spectroscopic tools available at synchrotron sources are much more diverse than those in individual laboratories, because the intense beams allow much more rapid data acquisition and because funding for and management of the facilities has allowed development of sophisticated multi-user devices. The websites of the major synchrotrons also provide useful introductions and information about their use (see Appendix A). Most of these facilities have permanent staff employed to assist new users.

Modern synchrotrons generate extremely high-intensity X-ray beams with large distributions of wavelengths (so-called synchrotron radiation) by bending circulating beams of positrons or electrons with powerful, superconducting magnets. The beam intensities of advanced (third-generation) synchrotrons are orders of magnitude greater than those of laboratory X-ray sources, allowing routine application of a broad range of previously inaccessible physical phenomena. For cement materials, diffraction methods, small angle scattering, X-ray absorption spectroscopy (XAS), and various imaging techniques have had the greatest impact. Specific studies include diffraction studies of phase development in hydrating cements; phase abundance in cement clinker and the structures of clinker phases; surface area and pore geometry investigations; imaging using soft-X-rays, X-ray absorption, and diffraction; and investigation of elemental structural environments using X-ray absorption spectroscopy. There are, however, many very effective synchrotron-based techniques that have yet to be used for cements. These include such surface-sensitive techniques as X-ray standing waves, X-ray reflectivity, grazing incidence X-ray absorption, infrared storage radiation (IRSR), and the synchrotron X-ray fluorescence microprobe, which provides spatially resolved major and trace element compositional information.

X-Ray Diffraction:
Crystal Structure and Phase Quantitation Studies

Because synchrotron beam intensities are very high, studies using diffraction techniques can be done much more rapidly than with laboratory sources and with much smaller samples. This high intensity allows, for instance, time-resolved studies of cement hydration, effective quantitative studies of clinker and cement mineralogy, and detailed studies of the structures of fine-grained crystalline phases. Both monochromatic and polychromatic X-rays are commonly used. The use of short-wavelength, high-energy X-rays (penetrating X-rays in the literature) offers another important advantage over laboratory X-ray sources, because the entire sample is analyzed, rather than just the few tens of µm near the surface, as in the standard $\theta/2\theta$ arrangement with Cu k_α radiation. The sample is held in a glass capillary tube and rotated, increasing particle statistics and reducing the effects of preferred orientation.

The structures of most crystalline phases relevant to cement-based materials have been investigated using standard laboratory X-ray sources, but as for other X-ray methods, synchrotron sources can offer significantly improved resolution and sensitivity and thus improved structural refinements. Yamazaki and Toraya[2] have used synchrotron powder diffraction to locate interlayer Ca and H_2O in an aluminous 1.1 nm tobermorite using Monte Carlo and Rietveld refinement of the structure previously determined by Merlino et al.[3] The results define the interlayer water positions, put the interlayer Ca atoms on partially occupied split positions with six-fold coordination, and locate the Al on the bridging tetrahedral sites. De la Torre et al.[4] used joint synchrotron and neutron powder diffraction to refine the structure of alite (MIII C_3S) containing small amounts of Mg and Al to simulate the alite of portland cement using Rietveld refinement methods. For the synchrotron work, they used a rotating sample in a glass capillary and short-wavelength [0.450294(6) Å] penetrating X-rays and were able to obtain high-resolution patterns in about 3 h. The neutron data used a longer wavelength (1.911 Å) and required 12 h of data acquisition. The final refined structure contains 155 structural sites (18 Si, 54 Ca, and 83 O) and captures the disorder by incorporating a large fraction of partially occupied Ca and O sites. The resulting Rietveld refinement provides good fits to the mineralogy of clinker and cement, and the quality of these fits does not depend greatly on Mg content of the alite. In related work, de la Torre et al.[5] discuss the problems related to obtaining quantitatively correct full phase analysis of clinker and cement using diffraction methods and show the clear advantages of synchrotron sources relative to laboratory sources. Peterson et al.[6] also demonstrate the effectiveness of synchrotron and neutron methods in clinker phase quantitation. Synchrotron powder X-ray data have also been used to refine the structure of thaumasite.[7]

There have been several in situ, time-resolved studies of the hydration of cement and related materials using synchrotron based X-ray diffraction techniques. Barnes et al.[8] have summarized these experiments. Again, the intensity of the synchrotron radiation offers significant advantages over laboratory sources, and either monochromatic X-rays or polychromatic X-rays (the energy-dispersive technique) can be used. In an early study, Clark and Barnes[9] showed that synchrotron energy-dispersive measurements for C_3S hydration provide more precise data than laboratory X-ray or neutron diffraction. The rate of decrease in intensity of the C_3S diffraction peaks is well fit by a linear Knudsen model and shows that D_2O greatly reduces the rate of reaction relative to H_2O. In a similar study, Jupe et al.[10] reported high-resolution, time-resolved data for C_3A hydration obtained with energy-dispersive penetrating synchrotron radiation. The sample was held in a spinning, cylindrical, 8 mm diameter polymer cell, and the diffractometer 2θ was set at 2.2°. The experimental arrangement allowed 5 and 0.3 s data acquisition times, and detectable changes in diffraction intensity were observed within 10–20 s of mixing the solid C_3A with water. The key to detection of this very rapid reaction was sampling of diffracted X-ray intensity from very small volumes, eliminating the decrease in temporal resolution related to slightly different reaction rates at different points in the reacting sample. Such studies are not practically possible using laboratory X-ray sources. Sample temperatures were increased by as much as 10°C above ambient (26°C) by beam heating. Initial reaction involved the formation of an intermediate phase (probably C_2AH_8) that can form and disappear within tens to hundreds of seconds. The final product was C_3AH_6; this phase did not begin to appear in significant amounts until the amount of intermediate product reached a maximum. These results, combined with other studies including gypsum in the reactants, show that the rate of formation of ettringite with sulfate ion present is even faster and prevents flash setting by preventing formation of the intermediate C_2AH_8.[11] Chang and Hou[12] reported similar results for slag cement hydration. Christensen et al.[13] reported a detailed study of cement and clinker hydration illustrating the improved resolution and sensitivity of monochromatic X-rays relative to energy-dispersive methods and also provide an excellent review of previous work. Shaw et al.[14-16] reported high-resolution results for in situ experiments studying the dehydration and hydrothermal formation of a number of Ca silicate phases.

X-Ray Absorption Spectroscopy

X-ray absorption spectroscopy is a powerful structural probe that can provide detailed atomic coordination numbers and interatomic distances for a wide variety of species in amorphous and crystalline materials and on solid surfaces. It analyzes the detailed shape and fine structure of X-ray absorption edges caused by the interference of incoming and outgoing X-rays as they scatter from atoms of the element of interest. In principle, XAS signals can be observed using laboratory X-ray sources, but in practice synchrotron sources are required for adequate signal to noise in almost all applications. There are two types of XAS analyses. XANES (X-ray absorption near edge structure) uses the shape of the absorption edge to fingerprint the structural environment by comparison with known materials. EXAFS (extended X-ray absorption fine structure) analyzes quantitatively the oscillations of the absorption intensity at energies greater than the absorption edge to yield coordination numbers and interatomic distances. Brown et al.[17] and Brown and Sturchio[18] provide useful introductions to XAS techniques and data analysis.

For cement systems, XAS has been used to study the Ca and Al environments in Ca aluminates and silicates and the environments of a wide variety of species of significance to nuclear and hazardous waste sequestration in C-S-H and hardened cement paste. Ca-XAS is important for cement-based materials, because there is no Ca isotope that can be routinely used for NMR studies, and understanding the local structural environments of Ca is essential for comprehensive structural understanding of disordered cement phases such as C-S-H. Richard et al.[19,20] used Ca- and Al-XAS to study several hydrous Ca aluminates, providing important structural insight into C_2AH_8 and CAH_{10} and their dehydration. For Ca silicates, Kirkpatrick et al.[21] and Lequeux et al.[22] used Ca-XAS to examine C-S-H and a number of crystalline phases. The results for the crystalline phases are for the most part in good agreement with structures known from X-ray diffraction. The data confirm the structural similarity of laboratory-synthesized C-S-H with C/S ratios from 0.7 to 1.4 to tobermorite rather than jennite. Comprehensive application of Ca-XAS to cement hydration has not been attempted, however, and lack of resolution in the Ca signal for different structural environments in, for example, C-S-H, AFm, AFt, and CH may limit its effectiveness for pastes.

For many minor species in cement, including many transition metals, XAS is the only effective, direct structural probe. There have been several important recent studies, and this technique should find a wide variety of future applications. In early work, Lee et al.[23] used Cr-XAS to show that Cr is present as toxic Cr^{6+} in OPC paste and that the addition of blast furnace slag reduces it to less toxic and less mobile Cr^{3+} with a significantly different local structure. Similarly, Allen et al.[24] have used Tc-XAS to investigate the effects of slag and other reducing agents to cement on the behavior of technetium, which is an abundant and very hazardous component of many nuclear wastes. The results show partial reduction of pertechnetate (TcO_4^-) by the slag and more extensive reduction by FeS, Na_2S, and $NaHPO_2$ additions. Structures similar to those of TcS_2 and TcO_2 were inferred.

More recently, there has been significant interest in understanding the behavior of potentially hazardous transition metals and several anionic species in C-S-H and cement paste, and XAS has played a central role in these studies. Rose et al.[25,26] have interpreted their Pb- and Zn-XAS data to show that the retarding effects of Pb^{2+} are due to its sorption on the surface of hydrating cement (mostly alite) grains and that Pb and Zn react with C-S-H to form Pb-O-Si and Zn-O-Si bonds. Tammaseo and Kersten[27] used Zn-, Ca-, and Si-XAS to reach the same conclusion concerning Zn-C-S-H and provide a detailed structural and thermodynamic model for this interaction. In contrast, Pomies et al.[28,29] used Cd-XAS and ^{113}Cd NMR to show that Cd^{2+} exchanges for Ca in C-S-H and that in hydrated C_3S it occurs in mixed Cd,Ca hydroxide, in C-S-H, and at high Cd concentrations as $Cd(OH)_2$. For Sn^{6+}, Bonhoure et al.[30] used Sn-XAS to suggest inner sphere sorption onto C-S-H in sorption experiments, but have proposed its incorporation in ettringite and formation of $CaZn(OH)_6$ in hardened cement pastes. Hsiao et al.[31] interpreted their Cu-XAS data for fly ash cements to indicate formation of $CuCl_2$, $Cu(OH)_2$, and Cu oxides as the origin of reduced Cu-leachability. Jing et al.[32] used As-XAS and Fourier transform IR spectroscopy to infer formation of Ca arsenates in cement-stabilized Fe hydroxide sludges. Bonhoure et al.[33] used I-XAS to show that C-S-H does not control the uptake of I^- and IO_3^- in cement paste.

Synchrotron-Based Imaging

Effective in situ imaging of cement, mortar, and concrete is a long-standing need, and synchrotron-based techniques such as soft X-ray microscopy, computed X-ray microtomography, and tomographic energy-dispersive diffraction imaging (TEDDI) are now coming on line and offer great potential. Of these techniques, soft X-ray microscopy has the greatest resolution (currently about 45 nm), but provides only a two-dimensional image for relatively thin samples. X-ray microtomography currently has a resolution of about 1 μm and produces three-dimensional images of samples up to several millimeters thick. TEDDI currently produces two-dimensional images showing the positions of specific crystalline phases with millimeter-scale resolution for centimeter-sized samples, but resolution may improve in the future.

The use of soft X-ray microscopy for in situ observation of hydrating cement systems was pioneered by Kurtis and coworkers[34–38] and has the potential to greatly improve understanding of the growth and morphologies of hydration products. It has not yet been applied to hydrated cement paste. The technique uses so-called zone plates as lenses to focus relatively low-energy (soft) X-rays with wavelengths in the 1–50 nm range onto the sample and produce an image on a CCD camera.[39] The samples are typically μL size (up to 40 μm thick) and may be wet or dry. They are held between silicon nitride windows and in situ observations can be made over extended times. The method has been used successfully for observation of C-S-H growth, the reaction of ASR gels with various solutions, and pozzolanic reactions of rice hull ash. For instance, Kurtis et al.[34,35] showed that ASR gels from in-service concrete react with alkali hydroxide solutions to form less dense gels (inferred to be due to silicate repolymerization) and with Ca hydroxide solutions to produce C-S-H with a spherulitic, sheaf of wheat morphology. This C-S-H morphology is also observed in the reaction of silica gel and silica fume with Ca hydroxide solution.[35,36] Hydration of β-

$(Ca,Ba)SiO_4$ with and without rice hull ash present produced a variety of more compact hydration product morphologies that depend on temperature and starting materials.[38]

Computed X-ray microtomography uses mono- or polychromatic X-rays to create three-dimensional images of the internal structures of objects using methods similar to conventional medical CAT scans, but with much greater spatial resolution (see Sutton et al.[40] for an introduction and Weiss et al.[41] and Butler et al.[42] for applications). The advantage of synchrotron radiation is that its high intensity and columnation allow enough radiation to penetrate the sample to provide sufficient signal for imaging of relatively dense objects such as cements and concretes. The fundamental approach is to measure many projections of some parameter through the object of interest at different angles and then reconstruct a three-dimensional image of that parameter. Transmission techniques produce images based on X-ray attenuation, diffraction techniques use the intensity of specific diffraction peaks, and absorption edge and fluorescence techniques use parameters related to elemental concentration. For instance, in their study of bone ingrowth by Ca-phosphate cements, Weiss et al.[41] used 635 transmission intensity images over an angular range of 180° with exposure times of 3–10 s per projection for objects about 1 mm on a side. The use of an optical microscope to observe the 2048 × 2048 pixel image of the CCD camera resulted in 1.4 µm resolution. Butler et al.[42] used transmission intensity to image pores in portland cement filled with toluene, demonstrating the potential of this technique in silicate systems. Similar studies have been undertaken with laboratory X-ray sources but at lower resolution.[43]

The TEDDI technique being developed by Hall and coworkers[44,45] uses imaging of diffraction intensity in the energy-dispersive mode to determine the location of different crystalline phases. In the initial work they were able to image calcite and dolomite in aggregate and portlandite and ettringite in the paste with approximately 0.5 mm resolution.

Absorption edge and X-ray fluorescence imaging have not been used for cement-based materials, but the examples for geological samples shown by Sutton et al.[40] demonstrate clearly the potential of these techniques to image compositionally different regions with a resolution of the order of one micrometer.

Neutron-Based Methods

Experimental probes of structure and dynamics based on neutron scattering are similar to some X-ray techniques, but in many cases provide otherwise unobtainable information. Neutron studies are especially important for cement-based materials, because neutrons scatter much more intensely from H atoms than do X-rays, allowing much more effective investigation of key issues related to water, OH groups, and pore structure than would otherwise be possible. The coherent neutron scattering cross for ^{2}H (deuterium) is also larger than that of ^{1}H. This allows effective structural studies of isotopically substituted samples. Bacon[46] and Squires[47] provide useful introductions to neutron diffraction and scattering, and the websites of many of the major neutron facilities also provide useful background (e.g., the NIST and LANSCE sites listed in Appendix B). Work of direct interest to cement and concrete science has focused on crystal structure investigations of individual phases, evaluation of the extent of cement hydration, investigations of the molecular-scale behavior of water in cement pastes, and investigations of surface area and pore structure in hydrated pastes. Neutron studies can involve use of elastic scattering (ENS), quasielastic scattering (QNS), and inelastic scattering (INS) by the material. ENS involves no net change in the energy (wavelength) of the scattered neutrons and is used in crystal structure studies using Bragg diffraction and small-angle scattering studies of larger-scale features in ways comparable to those traditionally used for X-rays. QNS used neutrons that have undergone relatively small changes in energy and for cements is used principally in studies of water mobility, allowing investigation of, for example, the amount of freely diffusing water in a hydrating cement sample. INS is a spectroscopic method that uses neutrons that have undergone significant change in their energy through interaction with atomic vibrations in the sample. It can, thus, probe the vibrational modes of crystalline or amorphous materials, including water. For crystalline phases, these are often analyzed in terms of the normal modes. Ghose[48] describes many types of studies using INS that have not been extensively applied to cement materials.

Neutron sources with fluxes adequate for structural studies of materials have been available since the 1950s. Like synchrotrons, these are principally national or international in scope, and many provide effective support for inexperienced users. Appendix B lists the websites of some of the major facilities worldwide. Large neutron fluxes are produced by nuclear reactors via controlled fission reactions and by spallation sources, in which high-energy particles, such as protons, from an accelerator bombard the nuclei of heavy elements, such as Hg, and knock neutrons from them. Since the 1950s, the fluxes available from reactor sources have remained nearly constant, and in the late 1980s and early 1990s, spallation sources such as LANSCE and ISIS began producing comparable fluxes. The SNS facility at

Oak Ridge National Laboratory is due to come on line in 2006 and will increase flux density by about two orders of magnitude. This spallation source should dramatically increase the application of neutrons in materials science, chemistry, and biology. Spallation sources produce pulses of neutrons, allowing improved time-of-flight measurements that provide wavelength-resolved data from all neutrons rather than only those in a small wavelength range.

Small-Angle Neutron Scattering and X-Ray Scattering Studies

Elastic scattering of neutrons and X-rays at small angles provides important and otherwise unobtainable information about surface area and the pore geometry of hydrated cement paste on the ~1–1 000 nm length scale. Indeed, these are the only effective tools available to characterize the in situ geometry of the paste pore structure. Small-angle neutron scattering (SANS) and X-ray scattering (SAXS) can be thought of as diffraction from regions that have coherent scattering length densities (mean atomic scattering length × atomic number density) significantly different from the bulk. Allen et al.[49] undertook the first comprehensive SANS study of hydrated cement paste and showed that its scattering comes from water (or D_2O) filled pores and thus that it is an effective tool for investigation of surface area and pore structure. They also provide a useful introduction to the experimental techniques and theory of data interpretation. In the small-angle region, the scattering intensity is usually evaluated as a function of the scattering vector

$$\mathbf{Q} = (4\pi / \lambda) \sin(\varphi_s / 2)$$

where λ is the neutron (or X-ray) wavelength and φ_s is the scattering angle. $\mathbf{Q}$ has units of length^{-1} and thus functions in reciprocal space. With current instruments, SANS data can be collected over a $\mathbf{Q}$ range from about 3×10^{-5} to 4 nm^{-1}. The size of the structural features probed by small-angle scattering is $2\pi/\mathbf{Q}$ and thus varies over several orders of magnitude, making SANS an especially effective probe of the fractal characteristics of cement paste microstructure. Modern SAXS instruments provide data over a $\mathbf{Q}$ range from about 10^{-2} to 1 nm^{-1}. In the large $\mathbf{Q}$ (Porod) region ($\mathbf{Q}R_g > 2.5$, where R_g is the radius of gyration of the scattering volumes), the intensity of the scattered neutrons $I = 2\pi |\Delta p|^2 S_v (1/\mathbf{Q}^4)$. Here S_v is the surface area per unit volume and Δp is the scattering contrast (here between the C-S-H and the water filled pores). Thomas et al.[50] have determined an accurate value for the scattering contrast of hydrated cement paste from data for samples with differing H_2O/D_2O ratios and showed that the interlayer water in C-S-H should be included in the C-S-H scattering density and not in the pores.

Using this value Thomas et al.[51] found surface areas for hydrated cement pastes in the range from about 140 to 190 m/cm^3 with the surface area increasing slightly with increasing water/cement (w/c) ratio. Using a model that calculates the density of D-dried paste, they found that the surface areas are generally quite similar to those obtained using BET gas absorption (~95–145 m^2/g). This area represents the surface of C-S-H (and other solid phases) in contact with all gel and capillary pores, but does not include the interfaces between unit-cell scale layers within C-S-H particles. Thomas et al.[51] show that their calculated surface area for samples hydrating in H_2O at 30°C correlates well with evolved heat up to about 26 h, but then remains approximately constant while the heat generation continues. Hydration in D_2O slows the reaction. The surface area per unit porosity at the start of hydration remains approximately constant. The authors interpret their data to support the presence of an early-formed, high–surface area (outer product) C-S-H and a later-formed, low–surface area (inner product) C-S-H. This interpretation is in good agreement with NMR relaxation studies of surface area development, but contrasts with older small-angle X-ray scattering data, which shows continued development of surface area up to values as large as 600 m^2/g after 1 year.[52,53] More recent ultra-small-angle X-ray scattering (USAXS) studies by Allen et al.[54] show that data at smaller **Q** values allows better evaluation of the integrated scattering intensity and leads to calculated surface area values of about 150 m^{2}cm^{-3} (95 m^{2}g^{-1} on a D-dried basis), in agreement with the range determined from SANS data. These authors argue that SANS and SAXS probe C-S-H and other surfaces that are related to features controlling permeability and rheology, whereas NMR studies and water sorprtion measurements, which yield much higher surface areas, provide information about the C-S-H gel.

In recent years, many workers have used fractal models of the hydrated paste microstructure to interpret SANS and SAXS data, and these models appear to describe well the scattering at **Q** values less than about 1 nm^{-1}. Features investigated by these data have direct space dimensions from about 1 nm to 1 μm, and most workers now interpret their data in terms of surface and mass (volume) fractals. Surface fractals are two-dimesional objects with topography that is self-similar (looks the same) at all length scales. The area of a surface fractal increases as the size of the probe used to measure it decreases, with the area following a power law relationship with probe size (here Q). The exponent is the fractal dimension. In the extreme, flat (Euclidian) surfaces have fractal dimensions of 2, and surfaces that are so rough that they fully fill space have fractal dimensions of 3. Mass (volume) fractals are three-dimensional objects with porosity that is self-similar at all length scales. Their density decreases as the volume sampled increases, with the mass scaling as a power of the volume. The exponent is the fractal dimension, and in the limit of a fully space filling (solid) object, the fractal dimension is 3. In most cases fractal objects follow a power law relationship over only a limited range of spatial scales, and this appears to be true for hydrated cement pastes. Schmidt,[55] Winslow et al.,[53] and Allen and Livingston[56] discuss the analysis of SANS and SAXS data in terms of fractal geometry.

Winslow et al.[52,53] were the first to obtain SAXS data for hydrated cement pastes over a wide enough range in Q space (~0.03–2 nm^{-1}) to allow discrimination of regions representing structures with different fractal dimensions. In the Q region representing features with direct space dimensions of about 3–20 nm, the data show objects with a mass fractal geometry that becomes more space filling with decreasing water saturation. At about 50% water saturation they become fully dense (fractal dimension = 3), and at lower saturations they take on surface fractal geometry. In contrast, features with characteristic lengths from about 20 to 200 nm have a rough surface fractal geometry (fractal dimensions of 2.3–2.9) at water saturations greater than 45%. At 40% water saturation these objects become fully dense (fractal dimension = 3). For oven-dried pastes, the geometries are quite different, with small-scale features having flat surface fractal geometry (fractal dimension = 2) and the large-scale features having mass fractal geometry. Winslow et al.[53] also showed that the geometry of the pastes does not depend on whether the degree of water saturation was reached by adsorption or desorption, as long as comparison is made at the same degree of saturation and not the same relative humidity.

Allen and Livingston[56] used SANS and ultra-small-angle X-ray scattering (Q = 0.3–0.004 nm^{-1}) to study hydrated cement pastes containing silica fume and also review earlier work by their group. They, too, interpret their data for wet pastes to indicate mass fractal geometry at small length scales and surface fractal geometry at larger scales. They attribute the surface fractal component to the surface decoration on the inner product C-S-H that replaces the cement grains and the mass fractal component to outer product C-S-H that fills space between the hydrating cement grains. They show that the fraction of surface area associated with the mass fractal (outer product) increases with increasing hydration time and that addition of silica fume retards this change. They observe an increase in total surface area in the blended samples relative to those without silica fume and attribute this to an increased amount of C-S-H due to pozzolanic reaction. Phair et al.[57] describe a combined SAXS and ultra-small-angle neutron scattering (USANS) study of alkali activated fly ash/kaolin, fly ash/slag, and OPC/slag mixtures that shows that the neutron scattering follows a fractal relationship, but that the models of Allen and coworkers for OPC and silica fume blends cannot be well applied in these cases. Phair et al.[57] discuss their data in terms of the size of large-scale scattering objects with dimensions of the order of micrometers, but quantitative values are difficult to obtain owing to multiple scattering effects. Qualitatively, the fly ash/slag mixtures have the smallest scattering particles. Popova et al.[58] have interpreted their SANS data for C-S-H formed by pozzolanic reaction of CaO and amorphous silica (Aerosil 200) to indicate a platelike C-S-H morphology with a maximum diameter of ~20 nm.

Structural interpretations based on the growing body of SANS/SAXS data for hydrated cement-based materials are now becoming an integral part of thinking about cement hydration. Jennings[59] has incorporated fractal characteristics into his comprehensive model of the structure of C-S-H in OPC, and Livingston[60] has developed a fractal-based nucleation and growth model for C_3S hydration. Fractal models of cement paste are likely to significantly influence advanced understanding of diffusion, creep, and other transport-related properties.

Neutron Diffraction Studies

Neutron diffraction studies are less common and generally more difficult than X-ray diffraction studies because neutron scattering is weaker for most elements than X-ray scattering. For cement systems, recent neutron diffraction work has focused on understand the freezing of pore water in hydrated paste, the internal stress imposed on the paste during freezing, Rietveld refinement of the structures of clinker phases, changes to the paste caused by heating, and the kinetics of cement hydration.

The behavior of concrete during freeze/thaw cycles is an important durability issue in cold regions, and in situ neutron diffraction of cement paste during low-temperature heating and cooling provides useful insight into the freezing and melting of water in the pore system. Fang et al.[61] demonstrated the formation of cubic ice (ice I_c) in a deuterated sample of hydrated paste. They observed no ice formation until 245 K during cooling (the same temperature at which freezing is observed by NMR for similar samples) and melting at above 273 K, but below the melting temperature of bulk D_2O of 277 K. Similar supercooling and hysteresis is known from many porous materials. In more carefully described studies, Schulson et al.[62] and Swainson and Schulson[63] show the formation of hexagonal ice (ice I_h, the stable ice phase at 1 atm and 273 K) during cooling of deuterated paste. Rather than a discrete freezing point, however, they observe significant crystallization during cooling beginning at 255 K and increasing crystallization to 230 K, the lowest temperature obtained. On heating they observe an approximately 25° hysteresis with a continuous decrease in the amount of ice up to about 277 K. They interpret their results in terms of freezing and melting of water in a distribution of pore sizes with crystallization occurring by growth of a percolating ice front rather than nucleation in individual pores. They interpret the observed broadening of the peaks for ice to be due to a correlation length for neutron scattering of about 8.5 nm, due possibly to ice I_h/I_c stacking disorder, small crystallites, or other defects. Swainson and Schulson[63] observe expulsion of approximately 23% of the water from a saturated paste (w/c = 0.4) at temperatures greater than 250 K (coincident with the first significant ice formation) and subsequent full reproducibility of the freeze/thaw behavior of the paste during two cycles to 230 K. They interpret the water loss to be driven by vapor transport owing to differences in the vapor pressure above solid and liquid water and suggest that these results indicate no significant change in the pore system during freezing. In a comment on this paper, Chatterji[64] suggests that the water loss caused significant microstructural change to the sample before the first freezing. In a related study, Schulson et al.[65] used neutron diffraction to monitor the changes in the unit cell dimensions of deuterated portlandite (CD) on cooling of cement paste to 20 K. The results show that differential thermal contraction between CD and C-S-H(D) causes increasing internal stress in hardened cement paste with decreasing temperature. The thermal anisotropy of CD causes compressive stress along crystallographic a (up to -40 ± 20 MPa and tensile stress along crystallographic c (30 ± 20 MPa).

Heating of concrete by, for instance, fire also causes substantial damage, and the compressive strength of typical concrete decreases abruptly at about 300°C. Castellote et al.[66] have used in situ neutron diffraction to study the thermally induced phase changes in 28-day cement pastes made with D_2O on heating up to 620°C and subsequent cooling in air. The results are correlated with subsequent Hg intrusion porosimetry and TGA. Neutron diffractions peaks for ettringite, portlandite (CD), calcite, C-S-H, and larnite (C_2S) are readily apparent for the initial sample. Peaks for ettringite begin to decrease by 50°C and disappear by 100°C. Loss of C-S-H intensity begins almost immediately, and C-S-H is not detectable above about 400°C for 8 × 40 mm cylindrical samples or above about 200°C for powdered samples. CD disappears at about 540°C, and CaO appears when CD disappears. On cooling, some of the lime rehydrates, but ettringite and C-S-H do not form. Integration of the background in the diffraction patterns provides an intensity parameter related to the total amount of H in the sample. Its evolution with increasing temperature reflects the observed phase changes. Differences in heating rate between 60 and 120°/h have little effect on the phase changes or porosity development, but examining the samples in powder form allows water and CO_2 loss at lower temperature than for the 8 × 40 mm cylinders. The MIP data show substantial increase in the amount of capillary porosity after the heating/cooling cycle. For the powdered sample, the authors attribute this entirely to mass loss, whereas for the cylindrical samples they believe that increased water vapor pressure also contributes to capillary porosity formation.

As discussed above under X-ray diffraction, quantitative analysis of the abundances of cement clinker phases in clinker or cement using diffraction requires excellent refinements of the structures of these phases, and synchrotron X-rays and neutrons have been used in these refinements. Berliner et al.[67] provide refinements for monoclinic and triclinic C_3S, β-C_2S, cubic and orthorhombic C_3A, and C_4AF based entirely on neutron data. De la Torre et al.[4] use both XRD and neutrons in a similar study, as described above.

Neutron diffraction has not been used extensively to study the development of crystalline phases during cement hydration. As described above, Clark and Barnes[9] compared the use of neutron, laboratory X-ray, and synchrotron X-ray sources to study cement hydration, and showed that neutron and laboratory X-ray sources give approximately the same analytical precision, but that synchrotron sources are better in this regard.

Quasielastic Neutron Studies

In QNS the motion of mobile atoms causes broadening of the energy range of the total elastic spectrum because of the addition or subtraction of energy from the scattered neutrons. Hydrogen has a very large incoherent scattering cross section and dominates the spectrum of cement materials. More than 99% of the signal from cured pastes is due to scattering from H.[68] The H atoms of mobile water with correlation times less than about 10^{-10} s cause significant peak broadening relative to those in OH groups or less-mobile water molecules. Thus, this method has been used principally to measure the amount of water consumed in hydration reactions and to probe the water dynamics in these systems. Typical spectra of hydrated paste clearly contain more than one component, but because the various components are centered at the same position, unique deconvolution is difficult. In early studies, FitzGerald et al.[68] and Berliner et al.[69] assumed that the elastic peak could be fit with a narrow component (a δ-function or narrow Gaussian) representing bound H and a broader, Lorentzian component representing free water. Recent work has shown that more complex procedures provide better fits. Thomas et al.[70] have proposed a three-component fit of data for hydrating C_3S and portland cement. The components represent structural H bound in hydrate phases (the narrow Gaussian component), "constrained" water (a narrow Lorentzian component), and free liquidlike water (a broader Lorentzian component). Use of this model is supported by the resolution of anomalies in the amount of bound H determined for the same sample at different temperatures. FitzGerald et al.[71] used a similar model for hydrating C_3S and portland cement and interpret the results to indicate that hydration at higher temperatures yields products that are more impervious to water ingress than those produced at lower temperatures during the diffusion-limited portion of the hydration process. In a different approach, Fratini and coworkers[72-74] have fit their QNS data for hydrating C_3S with and without retarder present with a narrow Gaussian representing "immobile" (bound) water and a stretched exponential representing "glassy" water. The stretched exponential term enters into the intermediate scattering function through a term $\exp-(t/\tau)^\beta$, where the stretching exponent, β, is a measure of the extent of interaction of the water with the surface. For bulk water, β = 1, and for increasing surface effects it takes on smaller values. Two different fitting routines using this concept provide acceptable fits to the data. Such stretched exponential functions are commonly used in curve fitting when combinations of Gausssian or Lorentizian functions do adequately match observations. Here, use of this function is based on arguments relating the dynamic behavior of water on solid surfaces to that of bulk, supercooled water.

For the available QNS data for hydrating cements, both the two- and the three-component procedures seem to adequately fit the observations. Additional work is needed to evaluate the range of applicability of these models and how well they describe the physical reality in hydrating cement systems. There are at least two issues that need resolution: How well do these procedures simulate the amount of bound water, and how well do they provide insight into the amount, microstructural environment, and dynamic behavior of the nonbound water that produces the broader spectral component?

Inelastic Neutron Studies

Spectroscopic studies of cement materials using INS are not as well developed as in condensed-matter physics and mineralogy, where detailed understanding of the phonon dispersion relations, the phonon density of states, or the magnon (magnetic spin wave) dispersion relations are more central. INS has been used by FitzGerald et al.[75] to monitor the amount of $Ca(OH)_2$ and residual mobile water (correlation times $\geq 10^{-10}$ s) during the hydration of C_3S. The amount of crystalline and noncrystalline $Ca(OH)_2$ was monitored as the intensity of the characteristic scattering band centered at 41 meV, which is due to motions of the OH groups linked to Ca. (There are also bands at 32 and 46 meV that are also associated with lattice modes in CH.) The spectra required about 35 min to collect, and were acquired at closely spaced intervals for the first few days and at longer intervals for several weeks. The results show that formation of CH is progressively delayed with decreasing temperature, but that the final amount of CH does not vary from 10 to 40°C. The results confirm previous observations that the acceleratory period of hydration precedes significant CH precipitation. The bulk of the data are well fit by an Avrami-type relationship, but at longer times the rate decreases, suggesting diffusion-limited reaction. Combining the data with previous QNS results for the amount of bound H allowed the author to calculate the amount of water in C-S-H and CH. The results were interpreted to show that the amount of H in C-S-H decreases with increasing temperature. Thomas et al.,[76] however, have used INS to show that Ca-OH linkages that yield INS spectral bands similar to those of CH are present in C-S-H, in agreement with phase equilibrium, NMR, and IR studies.[76-78] This observation suggests that FitzGerald et al.[75] may have underestimated the amount of water in their C-S-H by including it in CH. The results from Thomas et al.[76] also suggest that the concentration of Ca-OH linkages in C-S-H estimated by INS is the same as estimated from ^{29}Si MAS NMR data at high C/S ratios but is significantly higher at lower C/S ratios, near 1.0. If correct, this would suggest the coexistence of Ca-OH and Si-OH linkages in C-S-H in this compositional range.

Spectroscopic Probes of Molecular Scale Structure

In addition to the XAS techniques described above, there are many spectroscopic methods that can be effectively used to investigate the structures of cement materials at the molecular scale, to characterize clinker phases and hydrated cement paste, and to investigate the chemical reactions occurring in cement systems. Infrared (IR) spectroscopy measures the absorption of infrared light by the vibrational modes in a material, thus providing a direct probe of the atomic configurations present and a characteristic fingerprint of known phases. It is a traditional method in cement science with a history dating back many decades.[79] During the 1980s and 1990s, application of magic-angle spinning (MAS) nuclear magnetic resonance (NMR) spectroscopy using ^{29}Si, ^{27}Al, and other nuclei grew explosively, and these techniques are now widely used. IR and NMR will not be discussed here. Other techniques, such as Raman, electron-energy-loss, and X-ray photo electron spectroscopies, however, each have their strong points and could be more widely applied to cement materials than they have.

Raman spectroscopy, like IR, probes the vibrational modes of a material, but it is generally less sensitive than IR. The Raman signal arises from the energy gain or loss (frequency shift) of photons as they scatter from the vibrational modes (phonons) in the material. It can be used for both crystalline and amorphous phases. To overcome the low signal intensity, most Raman experiments use high-intensity lasers to provide the exciting radiation. Bensted[80] was the first to effectively use Raman spectroscopy to study cement hydration and showed that it could be used to quantitatively track the decrease in the amount of C_3S as hydration proceeded. Conjeaud and Boyer[81] published Raman microprobe spectra of clinker phases that are useful in tracking their disappearance during cement hydration. Tarrida et al.[82] published the first results concerning C_3S hydration using modern laser Raman methods. They obtained well-resolved spectra for CH and C_3S and assigned the bands observed for C_3S to Si-O stretching and O-Si-O bending modes. The spectra for their synthetic C-S-H samples are poorly resolved, showing principally bands due to Si-O-Si bending. They used the intensities of the C_3S and CH bands relative to those of an internal TiO_2 standard to quantify the extent of hydration. Kirkpatrick et al.[83] and Cong and Kirkpatrick[84] obtained high-resolution micro-Raman spectra of 11 and 14 Å tobermorite, jennite, and synthetic C-S-H with C/S ratios from 0.88 to 1.85. They assigned the observed bands to Si-O stretching of Si-tetrahedra with Q^1 and Q^2 polymerizations, Si-O-Si bending involving tetrahedra with the same polymerizations, internal deformation of Si-O tetrahedra, and complex motions involving Ca polyhedra. The data are in good agreement with a defect tobermorite silicate polymerization model for the C-S-H based principally on ^{29}Si NMR data. In their recent Raman and ESEM study of the water vapor hydration of oil field cements, Deng et al.[85] were able to identify alite, belite, C_4AF, and gypsum in the unhydrated cement and were able to track the formation of CH and ettringite and the carbonation of CH as reaction progressed. For 3-month samples they also observed a broad band that they assigned to Si-O-Si bending in C-S-H, but they were not able to observe Si-O stretching bands. The C-S-H of hydrated cement is much more disordered than many synthetic samples, and Raman methods appear to be relatively ineffective at identifying and characterizing it in situ. In other structural studies, Renaudin et al.[86] used micro-Raman in conjunction with single-crystal XRD to characterize the molecular scale structure and dynamics of nitrate AFm and its thermal dehydration products, and Deb et al.[87] used it to characterize the thermal breakdown of ettringite. Fukuda et al.[88] used it in conjunction with electron probe analysis to show that the melt formed during remelting of belite contains high concentrations of alkalis, Al, and Fe and that the melt has a structure similar to Ca aluminoferrite. Jallad et al.[89] used near-infrared Raman imaging mapping in an optical microscope to characterize the development and location of thaumasite in OPC mortar that had undergone sulfate attack. Spatial resolution was about 7 µm, and the results correlate well with SEM/EDX results. Raman spectroscopy has also been used extensively to study biocompatible phosphate and glass ionomer cements.

Electron energy loss spectroscopy (EELS) measures the energy changes to incident electrons as they scatter from electrons in different energy levels in a material. It is typically done with electron microprobes or microscopes. Richardson et al.[90] and Brydson et al.[91] used K-edge Al-EELS to show that Al enters the bridging tetrahedral sites in C-S-H, and may be present in hydrotalcitelike environments in some cement pastes.

X-ray photoelectron spectroscopy (XPS, also known as ESCA) measures the energies of electrons ejected from a sample because of interaction with incident X-rays. The data are typically analyzed in terms of the binding energy of the different electronic states of specific elements, for instance, Ca 2p or O 1s. The binding energies vary with local chemical bonding environments in a way analogous to NMR chemical shifts, and it is possible to use them to characterize the nearest neighbor coordination of many elements and even Si polymerization. Black et al.[92] have used XPS to characterize crystalline Ca-silicate hydrate phases and also review the early literature on its use to study cement hydration, which dates to the 1970s. They suggest that modern methods and improved data analysis involving both binding energies and the energies of Auger electrons may improve the potential of this technique for investigating cement pastes.

Acoustic Microscopy

Microscopic characterization using optical and electron methods is an essential component of many investigations of cement hydration, but recent and developing methods including the various X-ray microscopy techniques described above and scanning acoustic microscopy (SAM)[93-96] add new capabilities. SAM uses ultrasound in the 25 MHz–2 GHz range that is focused through a sapphire lens coupled to the sample by a fluid (typically water). It produces an image based on contrast in acoustic impedance and, importantly, can be used to produce images of internal structures in concrete. Current spatial resolution appears to be about 10 μm, and fractures and characteristics of the interfacial transition zone and aggregate reaction can be observed.

Appendix A:
Internet Websites for Major Synchrotron Facilities

ALS (USA): http://www-als.lbl.gov/
APS (USA): http://www.aps.anl.gov/
ELETTRA (Italy): http://www.elettra.trieste.it/
ESRF (France): http://www.esrf.fr/
HASYLAB (Germany): http://www-hasylab.desy.de/
MAXLab (Sweden): http://www.maxlab.lu.se/
NSLS (USA): http://www.nsls.bnl.gov/
SSRL (USA): http://www-ssrl.slac.stanford.edu/

Photon Factory (Japan): http://pfwww.kek.jp/
PSI (Switzerland): http://www.psi.ch/
Spring 8 (Japan): http://www.spring8.or.jp/
SRS (UK): http://www.srs.ac.uk/

Appendix B:
Internet Websites for Some Major Neutron Facilities

BENSC (Germany): http://www.hmi.de/bensc/
BNL (USA): http://neutrons.phy.bnl.gov/
ESS (Europe): http://www.neutron-eu.net/
FLNP (Russia): http://nfdfn.jinr.dubna.su/
HFIR (USA): http://neutrons.ornl.gov/
ILL (France): http://www.ill.fr/
IPNS (USA): http://www.pns.anl.gov/
ISIS (UK): http://www.isis.rl.ac.uk/
ISSP (Japan): http://www.issp.u-tokyo.ac.jp/
JAERI (Japan): http://rrsys.tokai.jaeri.go.jp/
KURRI (Japan): http://www.rri.kyoto-u.ac.jp/
LANSCE (USA): http://www.lansce.lanl.gov/
LLB (France): http://www-llb.cea.fr/
LNS (Switzerland): http://lns.web.psi.ch/
NFL (Sweden): http://www.studsvik.uu.se/
NIST (USA): http://rrdjazz.nist.gov/
NPMP (Canada): http://neutron.nrc-cnrc.gc.ca/
RISØ (Denmark): http://www.risoe.dk/
SNS (USA): http://www.sns.gov/

References

1. P. A. Fenter, M. L. Rivers, N. C. Sturchio, and S. R. Sutton, eds., *Applications of Synchrotron Radiation in Low-Temperature Geochemistry and Environmental Science*. Reviews in Mineralogy and Geochemistry, vol. 49. Mineralogical Society of America, Washington, D.C., 2002.
2. S. Yamazaki and H. Toraya, "Determinations of Positions of Zeolitic Calcium Atoms and Water Molecules in Hydrothermally Formed Aluminum-Substituted Tobermorite-1.1 nm Using Synchrotron Radiation Powder Diffraction Data," *J. Am. Ceram. Soc.*, **84**, 2685–2690 (2002).
3. S. Merlino, E. Bonaccorsi, and T. Armbruster, "Tobermorite: Their Real Structure and Order-Disorder (OD) Character," *Am. Mineral.*, **84**, 1613–1621 (1999).

4. A. G. de la Torre, S. Bruque, J. Campo, and M. A. G. Aranda, "The Superstructure of C_3S from Synchrotron and Neutron Powder Diffraction and Its Role in Quantitative Phase Analysis," *Cem. Concr. Res.*, **32**, 1347–1356 (2002).

5. A. G. de la Torre, A. Cabesa, A. Calvente, S. Bruque, and M. A. G. Aranda, "Full Phase Analysis of Portland Cement Clinker by Penetrating Synchrotron Powder Diffraction," *Anal. Chem.*, **73**, 151–156 (2001).

6. V. Peterson, B. Hunter, A. Ray, and L. P. Aldridge, "Rietveld Refinement of Neutron, Synchrotron, and Combined Powder Diffraction Data of Cement Clinker," *Appl. Phys. A*, **74** (suppl.), S1409–S1411 (2002).

7. S. J. Barnett, C. D. Adam, A. R. W. Jackson, and P. D. Hywel-Evans, "Identification and Characterization of Thaumasite by XRPD Techniques," *Cem. Concr. Res.*, **21**, 123–128 (1999).

8. P. Barnes, S. Colston, B. Craster, C. Hall, A. Jupe, S. Jacques, J. Cockroft, S. Morgan, M. Johnson, D. O'Connor, and M. Bellotto, "Time- and Space-Resolved Dynamic Studies on Ceramic and Cementitious Materials," *J. Synchrotron Radiation*, **7**, 167–177 (2000).

9. S. M. Clark and P. Barnes, "A Comparison of Laboratory, Synchrotron and Neutron Diffraction for the Real Time Study of Cement Hydration," *Cem. Concr. Res.*, **25**, 639–646 (1995).

10. A. C. Jupe, X. Turrillas, P. Barnes, S. L. Colston, C. Hall, D. Häusermann, and M. Hanfland, "Fast In Situ X-Ray Diffraction Studies of Chemical Reactions: A Synchrotron View of the Hydration of Tricalcium Aluminate," *Phys. Rev. B*, **53**, R14697–R14700 (1996).

11. P. Barnes, X. Turrillas, A. C. Jupe, S. L. Colston, D. O'Connor, R. J. Livesey, P. Livesey, C. Hall, D. Bates, and R. Dennis, "Applied Crystallography Solutions to Problems in Industrial Solid State Chemistry: Case Examples with Ceramics, Cements and Zeolites," *J. Chem. Faraday Trans.*, **92**, 2187–2196 (1996).

12. P.-K. Chang and W.-M. Hou, "A Study on the Hydration Properties of High Performance Slag Concrete Analyzed by SRA," *Cem. Concr. Res.*, **33**, 183–189 (2002).

13. A. N. Christensen, N. V. Y. Scarlett, I. C. Madsen, T. R. Jensen, and J. C. Hanson, "Real Time Study of Cement and Clinker Phases Hydration," *Dalton Trans.*, pp. 1529–1536 (2003).

14. S. Shaw, C. M. B. Henderson, and B. U. Komanschek, "Dehydration/Recrystallization Mechanisms, Energetics, and Kinetics of Hydrated Calcium Silicate Minerals: An In Situ TGA/DSC and Synchrotron Radiation SAXS/WAXS Study," *Chem. Geol.*, **167**, 141–159 (2000).

15. S. Shaw, C. M. B. Henderson, and S. M. Clark, "Hydrothermal Synthesis of Cement Phases: An In Situ Synchrotron, Energy Dispersive Diffraction Study of Reaction Kinetics and Mechanisms," *High Pressure Res.*, **20**, 311–321 (2001).

16. S. Shaw, C. M. B. Henderson, and S. M. Clark, "In-Situ Synchrotron Study of the Kinetics, Thermodynamics, and Reaction Mechanisms of the Hydrothermal Crystallization of Gyrolite, $Ca_{16}Si_{24}O_{60}(OH)_8 \cdot 14H_2O$, *Am. Mineral.*, **87**, 533–541 (2002).

17. G. E. Brown, G. Calas, G. A. Waychunas, and J. Petiau, "X-Ray Absorption Spectroscopy: Applications in Mineralogy and Geochemistry"; in *Spectroscopic Methods in Mineralogy and Geology*. Reviews in Mineralogy, vol. 18. Edited by F. C. Hawthorne. Mineralogical Society of America, Washington, D.C., 1988.

18. G. E. Brown and N. C. Sturchio, "An Overview of Synchrotron Radiation Applications to Low Temperature Geochemistry and Environmental Science"; in *Applications of Synchrotron Radiation in Low-Temperature Geochemistry and Environmental Science*. Reviews in Mineralogy and Geochemistry, vol. 49. Edited by P. A. Fenter, M. L. Rivers, N. C. Sturchio, and S. R. Sutton. Mineralogical Society of America, Washington D.C., 2002.

19. N. Richard, N. Lequeux, and P. Boch, "An EXAFS Study of Cementitious Phases," *Annal. Phys.* (Suppl.), **20**, 23–31 (1995).

20. N. Richard, N. Lequeux, and P. Boch, "EXAFS Study of Refractory Cement Phases: $CaAl_2O_{14}H_{20}$, $Ca_2Al_2O_{13}H_{16}$, $Ca_3Al_2O_{12}H_{12}$," *J. Phys. III France*, **5**, 1849–1864 (1995).

21. R. J. Kirkpatrick, G. E. Brown, N. Xu, and X.-D. Cong, "Ca X-ray Absorption Spectroscopy of C-S-H and Some Model Compounds," *Adv. Cem. Res.*, **9**, 31–36 (1997).

22. N. Lequeux, A. Morau, S. Philippot, and P. Boch, "Extended X-ray Absorption Fine Structure Investigation of Calcium Silicate Hydrates," *J. Am. Ceram. Soc.*, **82**, 1299–1306 (1999).

23. J. F. Lee, S. Bajt, S, B. Clark, G. M. Lamble, C. A. Langton, and L. Ojl, "Chromium Speciation in Hazardous, Cement-Based Waste Forms," *Physica B*, **209**, 577–578 (1995).

24. P. G. Allen, G. S. Siemering, D. K. Shuh, J. J. Bucher, N. M. Edelson, C. A. Langton, S. B. Clark, T. Reich, and M. A. Denecke, "Technetium Speciation in Cement Waste Forms Determined by X-Ray Absorption Fine Structure Spectroscopy," *Radiochimica Acta*, **76**, 77–86 (1997).

25. J. Rose, I. Moulin, J.-L. Hazemann, A. Maison, P. M. Bertsch, J.-Y. Bottero, F. Mosner, and C. Haehnel, "X-Ray Absorption Spectroscopy Study of Immobilization Processes for Heavy Metals in Calcium Silicate Hydrates: I. Case of Lead," *Langmuir*, **16**, 9900–9906 (2000).

26. J. Rose, I. Moulin, A. Maison, P. M. Bertsch, M. R. Weisner, J.-Y. Bottero, F. Mosner, and C. Haehnel, "X-Ray Absorption Spectroscopy Study of Immobilization Processes for Heavy Metals in Calcium Silicate Hydrates. 2. Zinc," *Langmuir*, **17**, 3658–3665 (2001).

27. C. E. Tammaseo and M. Kersten, "Aqueous Solubility Diagrams for Cementitious Waste Stabilization Systems. 3. Mechanisms of Zinc Immobilization by Calcium Silicate Hydrate," *Environ. Sci. Technol.*, **36**, 2919–2925 (2002).

28. M.-P. Pomies, N. Lequeux, and P. Boch, "Speciation of Cadmium in Cement Part I. Cd^{+2} Uptake by C-S-H," *Cem. Conc. Res.*, **31**, 563–579 (2001).

29. M.-P. Pomies, N. Lequeux, and P. Boch, "Speciation of Cadmium in Cement Part II. C_3S Hydration with Cd^{+2} Solution," *Cem. Conc. Res.*, **31**, 571–576 (2001).

30. I. Bonhoure, E. Wieland, A. M. Scheidegger, M. Ochs, and D. Kunz, "EXAFS Study of Sn(IV) Immobilization by Hardened Cement Paste and Calcium Silicate Hydrates," *Environ. Sci. Technol.*, **37**, 2184–2191 (2003).

31. M. C. Hsiao, H. P. Wang, and Y. W. Yang, "EXAFS and XANES Studies of Copper in a Solidified Fly Ash," *Environ. Sci. Technol.*, **35**, 2532–2535 (2001).

32. C.-Y. Jing, G. P. Korfiatis, and X.-G. Meng, "Immobilization Mechanisms of Arsenate in Iron Hydroxide Sludge Stabilized with Cement," *Environ. Sci. Technol.*, **37**, 5050–5056 (2003).

33. I. Bonhoure, A. M. Scheidegger, E. Wieland, and R. Dahn, "Iodine Species Uptake by Cement and CSH Studied by I K-Edge X-Ray Absorption Spectroscopy," *Radiochim. Acta,* **90**, 647–651 (2002).

34. K. E. Kurtis, P. J. M. Monteiro, J. T. Brown, and W. Meyer-Ilse, "Imaging of ASR Gel by Soft X-Ray Microscopy," *Cem. Concr. Res.,* **28**, 441–421 (1998).

35. K. E. Kurtis, P. J. M. Monteiro, J. T. Brown, and W. Meyer-Ilse, "High Resolution Transmission Soft X-Ray Microscopy of Deterioration Products Developed in Large Concrete Dams," *J. Microscopy,* **196**, 288–298 (1999).

36. E. M. Gartner, K. E. Kurtis, and P. J. M. Montiero, "Proposed Mechanism of C-S-H Growth Tested by Soft X-Ray Microscopy," *Cem. Concr. Res.,* **30**, 817–822 (2000).

37. K. E. Kurtis and P. J. M. Monteiro, "Chemical Additives to Control Expansion of Alkali-Silica Reaction Gel: Proposed Mechanisms of Control," *J. Mater. Sci.,* **38**, 2027–2036 (2003).

38. K. E. Kurtis and F. A. Rodrigues, "Early Age Hydration of Rice Hull Ash Cement Examined by Transmission Soft X-Ray Microscopy," *Cem. Concr. Res.,* **33**, 509–515 (2003).

39. M. Hettwer and D. Rudolph, "Fabrication of a Condenser Zone Plate for X-Ray Microscopy"; in *X-Ray Microscopy and Spectromicroscopy.* Edited by J. Thieme, G. Schmahl, E. Umbach, and D. Rudolph. Springer-Verlag, Heidelberg, 1998.

40. S. R. Sutton, P. M. Bertsch, M. Newville, M. Rivers, A. Lanzirotti, and P. Eng, "Microfluorescence and Microtomography Analyses of Heterogeneous Earth and Environmental Materials"; in *Applications of Synchrotron Radiation in Low-Temperature Geochemistry and Environmental Science.* Reviews in Mineralogy and Geochemistry, vol. 49. Edited by P. A. Fenter, M. L. Rivers, N. C. Sturchio, and S. R. Sutton. Mineralogical Society of America, Washington D.C., 2002.

41. P. Weiss, L. Obadia, D. Magne, X. Bourges, C. Rau, T. Weitamp, I. Khairoun, J. M. Bouler, D. Chappard, O. Gauthier, and G. Daculsi, "Synchrotron X-Ray Microtomography (on a Micron Scale) Provides Three-Dimensional Imaging Representation of Bone Ingrowth in Calcium Phosphate Biominerals," *Biomaterials,* **24**, 4591–4601 (2003).

42. L. G. Butler, J. W. Owens, F. K. Cartledge, R. L. Kurtz, G. R. Byerly, A. J. Wales, P. L. Bryant, E. F. Emery, B. Dowd, and X.-G. Xie, "Synchrotron X-Ray Microtomography, Electron Probe Analysis and NMR of Toluene Waste In Cement," *Environ. Sci. Technol.,* **34**, 3269–3275 (2000).

43. S. R. Stock, N. K. Naik, A. P. Wilkerson, and K. E. Kurtis, "X-Ray Microtomography (MicroCT) of the Progression of Sulfate Attack of Cement Paste," *Cem. Concr. Res.,* **32**, 1673–1675 (2002).

44. C. Hall, S. L. Colston, A. C. Jupe, S. D. M. Jacques, R. Livingston, A. O. A. Ramadan, A. W. Amde, and P. Barnes, "Non-Destructive Tomography Energy-Dispersive Imaging of the Interior of Bulk Concrete," *Cem. Concr. Res.,* **30**, 491–495 (2000).

45. P. Barnes, A. C. Jupe, S. D. M. Jacques, S. L. Colston, J. K. Cockcroft, D. Hooper, M. Betson, C. Hall, S. Bare, A. R. Rennie, J. Shannahan, M. A. Carter, W. D. Hoff, M. A. Wilson, and M. C. Phillipson, "Topographic Energy Dispersive Diffraction Imaging of Static and Dynamic Systems," *Nondestructive Testing and Evaluation,* **17**, 143–167 (2001).

46. G. E. Bacon, *Neutron Diffraction,* 3rd ed. Clarendon Press, Oxford, UK, 1975.

47. G. L. Squires, *Introduction to the Theory of Thermal Neutron Scattering.* Cambridge University Press, Cambridge, UK, 1978.

48. S. Ghose, "Inelastic Neutron Scattering"; in *Spectroscopic Methods in Mineralogy and Geology.* Reviews in Mineralogy, vol. 18. Edited by F. C. Hawthorne. Mineralogical Society of America, Washington, D.C., 1988.

49. A. J. Allen, C. G. Windsor, V. Rainey, D. Pearson, D. D. Double, and N. McN. Alford, "A Small Angle Neutron-Scattering Study of Cement Porosities," *J. Phys. D: Appl. Phys.,* **15**, 1817–1883 (1982).

50. J. J. Thomas, H. M. Jennings, and A. J. Allen, "Determination of the Neutron Scattering Contrast of Hydrated Portland Cement Paste Using H_2O/D_2O Exchange," *Adv. Cem. Based Mater.,* **7**, 119–122 (1998).

51. J. J. Thomas, H. M. Jennings, and A. J. Allen, "The Surface Area of Cement Paste as Measured by Neutron Scattering: Evidence for Two C-S-H Morphologies," *Cem. Concr. Res.,* **28**, 897–905 (1998).

52. D. Winslow, J. M. Bukowski, and J. F. Young, "The Early Evolution of the Surface of Hydrating Cement," *Cem. Concr. Res.,* **24**, 1025–1032 (1994).

53. D. Winslow, J. M. Bukowski, and J. F. Young, "The Fractal Arrangement of Hydrated Cement Paste," *Cem. Concr. Res.,* **25**, 147–156 (1995).

54. A. J. Allen, J. J. Thomas, and H. M. Jennings, "Composition and Density of Amorphous Calcium Silicate Hydrate in Cement Systems from Combined Neutron and X-Ray Small Angle Scattering," in preparation.

55. P. Schmidt, "Small-Angle Studies of Disordered, Porous, and Fractal Systems," *J. Appl. Cryst.,* **24**, 414–435 (1991).

56. A. J. Allen and R. A. Livingston, "Relationship between Differences in Silica Fume Additives and Fine-Scale Microstructural Evolution in Cement Based Materials," *Adv. Cem. Based. Mater.,* **8**, 118–131 (1998).

57. J. W. Phair, J. C. Schulz, W. K. Bertram, and L. P. Aldridge, "Investigation of the Microstructure of Alkali-Activated Cements by Neutron Scattering," *Cem. Concr. Res.,* **33**, 1811–1824 (2003).

58. A. Popova, G. Geoffroy, E. M. Gartner, and A. Lapp, "Calcium Silicate Hydrates Studied by Small-Angle Neutron Scattering (SANS)," *J. Am. Ceram. Soc.,* **85**, 1303–1305 (2002).

59. H. M. Jennings, "A Model for the Microstructure of Calcium Silicate Hydrate in Cement Paste," *Cem. Concr. Res.,* **30**, 101–116 (2000).

60. R. A. Livingston, "Fractal Nucleation and Growth Model for the Hydration of Tricalcium Silicate," *Cem. Concr. Res.,* **30**, 1853–1860 (2002).

61. M. P. Fang, P. E. Sokol, J. Y. Jehng, and W. P. Halperin, "Neutron Diffraction Study of Cement," *J. Porous Mater.,* **6**, 95–99 (1999).

62. E. M. Schulson, I. P. Swainson, T. M. Holden, and C. J. Korhnon, "Hexagonal Ice in Hardened Cement," *Cem. Concr. Res.,* **30**, 191–196 (2000).

63. I. P. Swainson and E. M. Schulson, "A Neutron Diffraction Study of Ice and Water within a Hardened Cement Paste During Freeze-Thaw," *Cem. Concr. Res.,* **31**, 1821–1830 (2001).

64. S. Chatterji, "A Discussion of the Paper 'A Neutron Diffraction Study of Ice and Water within a Hardened Cement Paste During Freeze-Thaw,'" *Cem. Concr. Res.,* **32**, 1845–1846 (2002).

65. E. M. Schulson, I. P. Swainson, and T. M. Holden, "Internal Stress within Hardened Cement Paste Induced through Thermal Mismatch," *Cem. Concr. Res.,* **31**, 1785–1791 (2001).

66. M. Castellote, C. Alonso, C. Andrade, X. Turrillas, and J. Campo, "Composition and Microstructural Changes of Cement Pastes upon Heating, as Studied by Neutron Diffraction," *Cem. Concr. Res.,* in press (2004).

67. R. Berliner, C. Ball, and P. B. West, "Neutron Powder Diffraction Investigation of Model Cement Compounds," *Cem. Concr. Res.,* **27**, 551–575 (1997).

68. S. A. FitzGerald, D. A. Neumann, and J. J. Rush, "In Situ Quasi-Elastic Neutron Scattering Study of the Hydration of Tricalcium Silicate," *Chem. Mater.,* **10**, 397–402 (1998).

69. R. Berliner, M. Popovici, K. Herwig, H. M. Jennings, and J. Thomas, "Neutron Scattering Studies of Hydrating Cement Systems," *Physica B,* **241–243**, 1237–1239 (1998).

70. J. J. Thomas, S. A. FitzGerald, D. A. Neumann, and R. A. Livingston, "State of Water in Hydrating Tricalcium Silicate and Portland Cement Pastes as Measured by Quasi-Elastic Neutron Scattering," *J. Am. Ceram. Soc.,* **84**, 1811–1816 (2001).

71. S. A. FitzGerald, J. J. Thomas, D. A. Neumann, and R. A. Livingston, "A Neutron Scattering Study of the Role of Diffusion in the Hydration of Tricalcium Silicate," *Cem. Concr. Res.,* **32**, 409–413 (2002).

72. E. Fratini, S.-H. Chen, P. Baglioni, and M.-C. Bellissent-Funel, "Age-Dependent Dynamics of Water in Hydrated Cement Paste," *Phys. Rev. E,* **64** [020201(R)] 1–4 (2001).

73. E. Fratini, S.-H. Chen, P. Baglioni, and M.-C. Bellissent-Funel, "Quasi-Elastic Neutron Scattering Study of Transitional Dynamics of Hydration Water in Tricalcium Silicate," *J. Phys. Chem., B,* **106**, 158–166 (2002).

74. P. Baglioni, E. Fratini, and S.-H. Chen, "Glassy Dynamics of Water in Hydrated Cement Paste," *Appl. Phys. A* (Suppl.), **74**, S1178–S1181 (2002).

75. S. A. FitzGerald, D. A. Neumann, J. J., Rush, R. J. Kirkpatrick, X.-D. Cong, and R. A. Livingston, "Inelastic Neutron Scattering Study of the Hydration of Tricalcium Silicate," *J. Mater. Res.,* **14**, 1160–1165 (1999).

76. J. J. Thomas, J. J. Chen, H. M. Jennings, and D. A. Neumann, "Ca-OH Bonding in the C-S-H Gel Phase of Tricalcium Silicate and White Portland Cement Pastes Measured by Inelastic Neutron Scattering," *Chem. Mater.,* **15**, 3813–3817 (2003).

77. X.-D. Cong and R. J. Kirkpatrick, "^{17}O MAS NMR Investigation of the Structure of Calcium Silicate Hydrate Gel," *J. Am. Ceram. Soc.,* **79**, 1585–1592 (1996).

78. P. Yu, R. J. Kirkpatrick, B. Poe, P. McMillan, and X.-D. Cong, "Structure of Calcium Silicate Hydrate (C-S-H): Near-, Mid- and Far-Infrared Spectroscopy," *J. Am. Ceram. Soc.,* **82**, 742–748 (1999).

79. H. F. W. Taylor, *Cement Chemistry,* 2nd ed. Telford, London, 1997.

80. J. Bensted, "Uses of Raman Spectroscopy in Cement Chemistry," *J. Am. Ceram. Soc.,* **59**, 140–143 (1976).

81. M. Conjeaud and H. Boyer, "Some Possibilities of Raman Microprobe in Cement Chemistry," *Cem. Concr. Res.,* **10**, 61–70 (1980).

82. M. Tarrida, M. Madon, B. LeRolland, and P. Colombet, "An In-Situ Raman Spectroscopy Study of the Hydration of Tricalcium Silicate," *Adv. Cem. Based Mat.,* **2**, 15–20 (1995).

83. R. J. Kirkpatrick, J. L. Yarger, P. F. McMillan, P. Yu, and X.-D. Cong, "Raman Spectroscopy of C-S-H, Tobermorite and Jennite," *Adv. Cem. Based Mater.*, **5**, 93–99 (1997).
84. X.-D. Cong and R. J. Kirkpatrick, "The Structure of C-S-H: Review of Spectroscopic Data"; pp. 143–158 in *Proceedings of the Second International Conference on NMR of Cements.* Edited by P. Colombet et al. Springer-Verlag, 1998.
85. C.-S. Deng, C. Breen, J. Yarwood, S. Habesch, J. Phipps, B. Craster, and G. Maitland, "Aging of Oilfield Cement at High Humidity: A Combined FEG-ESEM and Raman Microscopic Investigation," *J. Mater. Chem.*, **12**, 3105–3112 (2002).
86. G. Renaudin, J.-P. Rapin, B. Humbert, and M. Francois, "Thermal Behavior of the Nitrated AFm Phase $Ca_4Al_2(OH)_{12}(NO_3)_2 \cdot 4H_2O$," *Cem. Concr. Res.*, **30**, 307–314 (2000).
87. S. K. Deb, M. H. Manghnani, K. Ross, R. A. Livingston, and P. J. M. Montero, *Phys. Chem. Minerals*, **30**, 31–38 (2003).
88. K. Fukuda, A. Takeda, A. Yamaguchi, S. Hashimoto, and H. Yoshida, "Characterization of Liquid Exsolved by Remelting Reaction of Belite," *J. Am. Ceram. Soc*, **84**, 1155–1160 (2001).
89. K. N. Jallad, M. Santhanam, M. D. Cohen, and D. Ben-Amotz, "Chemical Mapping of Thumasite Formed in Sulfate-Attacked Cement Mortar Using Near-Infrared Raman Imaging Microscopy," *Cem. Concr. Res.*, **31**, 953–958 (2001).
90. I. G. Richardson, A. R. Brough, R. Brydson, G. W. Groves, and C. M. Dobson, "Location of Aluminum in Substituted Calcium Silicate Hydrate (C-S-H) Gels as Determined by Si-29 and Al-27 NMR and EELS," *J. Am. Ceram. Soc.*, **76**, 2285–2288 (1993).
91. R. Brydson, I. G. Richardson, and G. W. Groves, "Determining the Local Coordination of Aluminum in Cement Using Electron-Energy-Loss Near-Edge Structure," *Mikrochim. Acta*, **114**, 221–229 (1994).
92. L. Black, K. Garbev, P. Stemmermann, K. R. Hallam, and G. C. Allen, "Characterization of Crystalline C-S-H Phases by X-Ray Photoelectron Spectroscopy," *Cem. Concr. Res.*, **33**, 899–911 (2003).
93. R. A. Livingston, M. H. Manghnani, and M. Prasad, "Characterization of Portland Cement Concrete Microstructure Using the Scanning Acoustic Microscope," *Cem. Concr. Res.*, **29**, 287–291 (1999).
94. P. V. Zinin, M. H. Manghnani, Y. C. Wang, and R. A. Livingston, "Detection of Cracks in Concrete Composites Using Acoustic Microscopy," *Nondest. Testing Eng. Inter.*, **33**, 283–287 (2000).
95. P. V. Zinin, M. H. Manghnani, C. Newtson, and R. A. Livingston, "Acoustic Microscopy Imaging of Steel Reinforcement in Concrete," *J. Nondest. Eval.*, **21**, 105–110 (2002).
96. M. Prasad, M. H. Manghnani, Y. Wang, P. V. Zinin, and R. A. Livingston, "Acoustic Microscopy of Portland Cement Mortar Aggregate/Paste Interfaces," *J. Mater. Sci.*, **35**, 3607–3613 (2000).

Influence of Initial and Exposure Conditions on Moisture Transport in Hydrated Cement Systems

Y. Maltais, J. Marchand, K. Hazrati, and M. Jolin

The results of an investigation carried out to quantify the influence of initial and boundary conditions on moisture transport in cement-based materials are presented in this chapter. The test program was designed to establish the influence of various boundary and initial conditions on the flux of moisture through different mortar slabs. The influence of boundary and initial conditions on both water absorption (or capillary suction) and water vapor adsorption was also investigated using well-hydrated mortar specimens. Finally, the test program was designed to determine if the presence of sodium chloride (in the material pore solution) has any influence on both water vapor sorption isotherms and drying kinetic of cement-based materials. Experimental work was carried out on well-hydrated mortar and cement paste samples prepared at 0.40 and 0.60 water/cement ratios using two standardized Canadian cements (CSA Type 10 and CSA Type 50). Water vapor transmission, capillary suction, water vapor adsorption, and desorption tests were also carried out. Results obtained in this investigation clearly identified the influence of boundary and initial conditions on moisture flux measured during water vapor transmission tests. From a kinetic point of view, test results have shown that there is an important difference between liquid water absorption and water vapor adsorption (at 100% RH). However, from a thermodynamic point of view, test results obtained upon the completion of absorption and adsorption tests are quite similar. Finally, test results have established that both equilibrium water content and drying kinetics are significantly influenced by the presence of sodium chloride salt in the material's porous structure.

Introduction

Moisture transport plays an important role in the degradation of many types of concrete structures. For instance, most concrete structures are not maintained in a saturated state, and moisture transport near the exterior surfaces is often considered to have a relatively important influence on the penetration of aggressive species.[1] This is true for concrete structures in contact with seawater and deicing salts that often suffer from premature degradation by reinforcing steel corrosion and chemical attack.[2] This is also the case for concrete slabs and foundations exposed to sulfate-laden soils that may sustain premature degradation if improperly designed.[3] In the latter case, the excessive transport of moisture (containing sulfates) in porous concrete mixtures is often found to foster degradation by salt weathering.[4–7]

As emphasized by Surprenant and Malish,[8] the transport of moisture also has a direct impact on construction practices and the durability of building components placed in contact with concrete. For instance, the placement of floor coverings (such as carpets and linoleum) is often delayed to allow concrete to sufficiently dry.[9,10] The penetration of excessive moisture in residential and commercial buildings is also known to cause numerous problems ranging from tile delamination and wood-panel discoloration to the growth of mold and mildew.

Like all transfer phenomena, the transport of moisture in concrete depends on the initial and boundary conditions. Indeed, as emphasized by Nilsson et al.[11] and Hazrati,[12] the flux of moisture for a given mixture is influenced, to various degrees, by the gradient in relative humidity across the material, the initial water content of the solid, the ambient temperature, and the chemical composition of the pore solution. Over the last decade, numerous papers have been published on the mechanisms of water transport in cementitious materials.[13-16] Unfortunately, experimental data on the effect of boundary and initial conditions on moisture transport in cement-based materials remain relatively limited.

Scope of the Investigation

The results of a comprehensive investigation carried out to quantify the influence of initial and boundary conditions on moisture transport in mature mortar mixtures are reviewed in this chapter. The test program was designed to provide quantitative information on the effects of exposure conditions on various phenomena, such as water absorption (or capillary suction) and water vapor adsorption. Tests were also run to verify if the presence of sodium chloride (in the material pore solution) and the type of sample conditioning treatment had any influence on both the water vapor sorption isotherms and the drying kinetics of hydrated cement systems.

It should be emphasized that this investigation was greatly inspired by the pioneering work of Brewer,[17] who published half a century ago the results of a comprehensive study of the mechanisms of moisture transport in concrete. Results generated by Brewer provided useful information on the influence of boundary conditions on the kinetics of moisture transports. Brewer's data are still commonly cited today in papers dealing with the mechanisms of water transport in concrete.[8] However, Brewer did not address in his 1965 study the influence of initial conditions on the transport of moisture. In that respect, the present work is highly complementary to Brewer's contribution.

Overall, the results of this study should be of interest to scientists and engineers concerned with the transport of moisture and ions in hydrated cement systems. It should also provide useful data to those investigating the transport of water in concrete using standard procedures such as the moisture dome test (ASTM E 1907, ASTM F 1869) and the moisture transmission test (ASTM E 96, ISO 12572).

Test Program

Four test series were performed in this investigation. In the first one, water vapor transmission tests were run on well-hydrated mortar slabs (nominal thickness = 25 mm). Test variables included water/cement (w/c) ratio (0.40 and 0.60), initial material water content (initially dry or saturated), and the nature of boundary conditions at the sample's bottom surface (direct contact with a free water surface or contact with a water vapor saturated environment — 100% relative humidity [RH]). In all cases, the top surface of the samples was exposed to a low relative humidity environment (maintained around 5% RH) using an absorbing desiccant (silica gel). The mass gain of the gel was regularly measured and renewed throughout the entire duration of the test.

In the second series, the uptake of water by capillary suction and vapor adsorption was investigated. Test results were systematically compared to establish if the nature of boundary conditions (i.e., direct contact with free water surface or contact with a water vapor saturated environment) has a significant influence on the kinetics of moisture migration. Tests were carried out on well-hydrated mortar samples ($35 \times 35 \times 30$ mm) produced at two different water/cement ratios (0.40 and 0.60) and that had been conditioned for several months at different relative humidity levels ranging from 33 to 85%.

The third test series was carried out to determine if the presence of sodium chloride (in the material pore solution) had any influence on the drying kinetics of the mortar samples (w/c = 0.40 and 0.60). The dry mortar samples ($35 \times 35 \times 30$ mm) were immersed in various sodium chloride solutions (prepared at 0, 1000, and 2000 mmol/L) for a 3-month period. After the immersion period, samples were exposed to a 75% RH environment for 120 days.

A final test series was performed to establish if the equilibrium water content in cement-based materials (usually given by water sorption isotherms) is influenced by the presence of sodium chloride in the material porous structure. Water sorption isotherms were determined using porous

Table I. Chemical composition of portland cements and standardized crushed siliceous sand

Oxides (%)	CSA T10		CSA T50	Sand (%)
	Series 1, 3, 4	Series 2*	Series 4	
SiO_2	19.78	20.72	21.45	99.71
Al_2O_3	4.39	4.15	3.58	0.122
TiO_2	0.22	0.20	0.21	0.012
Fe_2O_3	3.00	3.02	4.38	0.028
CaO	62.04	62.15	63.93	0.008
SrO	0.26	0.26	0.07	
MgO	2.84	2.53	1.81	0.005
Mn_2O_3	0.04	0.06	0.05	
Na_2O	0.32	0.19	0.24	0.003
K_2O	0.91	0.90	0.70	0.015
SO_3	3.20	3.43	2.28	
LOI	2.41	2.36	0.86	

*Portland cement used by Hazrati.[12]

Vycor glass and neat cement pastes (prepared at 0.60 w/c using two portland cements). These materials were immersed in sodium chloride solutions prepared at 0, 250, 500, 1000, and 2000 mmol/L. Porous Vycor glass was used to ensure that any influence of sodium chloride on the water sorption isotherms would not eventually be attributed to some alteration of the material's microstructure due, for example, to the formation of chloride-bearing phases.

Materials, Mixture Preparation, and Experimental Procedures

Materials and Mixture Characteristics

Mortar mixtures were prepared at two different water/cement ratios: 0.40 and 0.60. The sand volume fraction was fixed at 50%. All mortar mixtures were produced using a CSA Type 10 portland cement (ASTM Type I equivalent) and a standardized crushed siliceous sand (Ottawa sand) with a density of 2.60. Water vapor sorption isotherms were determined using porous vycor glass and neat cement pastes. Cement pastes were cast at 0.60 w/c

Table II. Mineralogical composition of portland cements

Compound (%)	CSA T10		CSA T50
	Series 1, 3, 4	Series 2*	Series 4
C_3S	59	54	62
C_2S	12	19	16
C_3A	7	6	2
C_4AF	9	9	13

*Portland cement used by Hazrati.[12]

Table III. Sand grain size distribution

Sieve (ASTM E-11)	% passing
20	100
30	100
40	72
50	30
60	17
100	1

Sand density = 2.60; sand absorption = 0.2%.

using CSA Types 10 and 50 portland cements (ASTM Types I and V equivalent). The chemical and mineralogical composition of the cements and sand are given in Tables I and II, respectively. The grain size distribution of the sand is presented in Table III.

All mixtures were prepared using deionized water. The mixtures were batched in a high-speed mixer placed under vacuum (at 10 mbar) to prevent, as much as possible, the formation of air voids during mixing. For the first, second, and fourth test series, mixtures were cast in plastic cylinders (diameter = 70 mm; height = 200 mm). The molds were sealed and rotated for the first 24 h in order to prevent bleeding. In addition, prismatic samples (200 × 35 × 35 mm) were cast in plastic molds and cured for 24 h under damp cloths. At the end of the initial curing period, cylinders and bars were all demolded and sealed with an aluminum adhesive foil for a minimum of 12 months. During this period, samples were maintained at room temperature.

Mortar samples used in the second test series were prepared under vacuum by Hazrati.[12] The prismatic samples (200 × 35 × 35 mm) were cast in plastic molds and cured for 24 h under damp cloths. Samples were then demolded and immersed in a saturated lime solution at room temperature for a minimum period of 1 year. Companion (50 × 50 × 50 mm) cubic specimens were also cast for compressive strength measurements. The characteristics of these mixtures and their average compressive strength test results (obtained by Hazrati[12]) are summarized in Table IV.

Table IV. Mixture characteristics and 28-day compressive strength test results

Identification	Material	w/c	Cement (kg/m³)	Water (kg/m³)	Sand (kg/m³)	Compressive strength* (MPa)
M10 - 0.40	Mortar	0.40	697	279	1325	41.5
M10 - 0.60	Mortar	0.60	545	327	1325	71.0
P10 - 0.60	Paste	0.60	1090	655	0	
P50 - 0.60	Paste	0.60	1090	655	0	

*28-day test results obtained by Hazrati.[12]

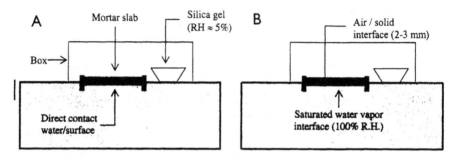

Figure 1. Experimental setup for the water vapor transmission tests: (A) direct contact between free water surface and mortar sample; (B) saturated water vapor air in contact with mortar sample.

Experimental Procedures

Water Vapor Transmission Tests (Test Series 1)

At the end of the curing period, mortar cylinders were cut with a small deionized water-cooled diamond saw. Samples were sawn from the central part of the cylinders. Mortar discs (25 mm thick each) were either oven dried (at 105°C for 5–7 days) or vacuum saturated (for a 3-day period). After the initial conditioning treatment, samples were mounted on moisture transmission cells similar to those shown in Fig. 1. As can be seen in the figure, the lower part of the samples was placed in direct contact with water or in contact with a saturated water vapor environment. The top part of the sample was exposed to an environment maintained around 5% relative humidity using silica gel. During the tests, the silica gel was weighed on a daily basis by opening the box for a few seconds to allow the measurement and the replacement of silica gel. Using a control box, the daily weight gain

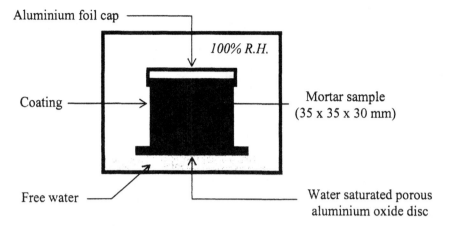

Figure 2. Experimental setup for the water absorption tests.

of silica gel was corrected to account for the water vapor contained in the air. During the water vapor transmission tests, pure drying experiments were also carried out. For those tests, mortar samples were coated on all faces except the top surface. Once coated, the samples were placed in a small container maintained at low relative humidity using silica gel. The silica gel was renewed on a daily basis.

Water Absorption and Water Vapor Adsorption Tests (Test Series 2)

In order to investigate the influence of initial and boundary conditions on water absorption and adsorption mechanisms, mortar samples conditioned for 24 months at 33, 59, and 85% relative humidity were used. The conditioning of samples at different relative humidity levels was carried out on mortar samples (35 × 35 × 30 mm) that had never undergone any drying pretreatment. During the 24-month conditioning period, samples were kept at 23°C and regularly flushed with high-purity nitrogen gas to avoid carbonation.

For the water absorption tests (also called capillary suction tests), mortar samples were placed on a coarse porous aluminum oxide disc, which was continuously kept saturated. In order to obtain a one-dimensional moisture flow, the four sides of the sample were coated. The upper side of the specimens was covered with an aluminum foil cap (see Fig. 2). During testing, mortar samples were weighted on an analog scale (±0.05 mg) at regular intervals from the onset of the water absorption tests. The absorption surface was wiped with a moist cloth before weighting. Mortar sample mass

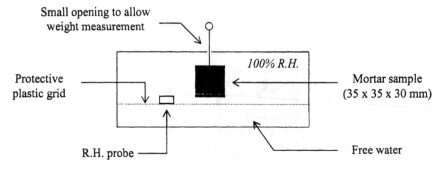

Figure 3. Experimental setup for the water adsorption tests.

gain was monitored for nearly 150 days. At the end of each test, mortar samples were placed under vacuum for a 48-hour period to measure the total water content (θ_{vac}).

Water vapor adsorption tests (carried out at 100% RH) were performed using the experimental setup shown in Fig. 3. For these tests, mortar samples were placed over a free deionized water surface. The relative humidity within the box reached nearly 100% within 1 h (see Fig. 4). As for capillary suc-

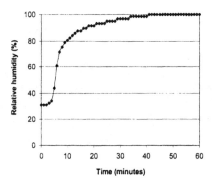

Figure 4. Relative humidity in a water adsorption test box.

tion tests, the top and lateral sides of the samples were coated. Mortar samples were weighted on an analog scale (±0.05 mg) at regular intervals and the mass gain was monitored for more than 150 days. At the end of each test, mortar samples were placed under vacuum for a 48-h period to obtain the total water content of the material (θ_{vac}).

Drying Experiments (Test Series 3)

Samples (35 × 35 × 30 mm) were first oven dried at 105°C to a constant weight. After the drying period, the specimens were immersed in sodium chloride solutions prepared at 0, 1000, and 2000 mmol/L for a 3-month period. After the immersion period, samples were weighted and placed in a sealed chamber similar to that shown in Fig. 3. The environment was maintained at a 75% RH using a saturated sodium chloride solution. All chambers were kept at 23°C and flushed every week with high-purity nitrogen gas to avoid carbonation. Samples were weighed on an analog scale (±0.05 mg) at regular intervals and mass loss was monitored for more than 150 days.

Table V. Relative humidity over salt saturated solutions at 23°C

Salt	Relative humidity (%)	
	Theoretical	Measured
KNO_3	94.6	94.0
KCl	85.1	86.4
$(NH_4)_2SO_4$	80.1	81.8
$SrCl_2 \cdot 6H_2O$	71.5	72.7
$Mg(NO_3)_2 \cdot 6H_2O$	53.5	56.5
K_2CO_3	44.0	46.0
$MgCl_2 \cdot 6H_2O$	33.1	32.5
$C_2H_3O_2K$	22.8	22.0
LiCl	11.3	10.3

Determination of Sorption Isotherms (Test Series 4)

Well-hydrated neat cement paste discs (diameter = 95 mm; thickness = 25 mm) were initially oven dried at 105°C to a constant weight. After this drying period, paste samples were immersed (for a 3-month period) in sodium chloride solutions prepared at 0, 1000, 2000, and 4000 mmol/L. Once the immersion period over, paste samples were crushed to particle size below 2 mm. Crushing took place in a moist room to avoid any drying of the samples. Wet powdered samples (approximately 10 g each) were then weighed and placed in nine different sealed conditioning chambers above different salt saturated solutions, which kept the relative humidity constant. The salts used for the tests and their corresponding relative humidities are listed in Table V. All chambers were kept at 23°C and were flushed every week with high-purity nitrogen gas to avoid sample carbonation. The equilibrium moisture content of the specimens was measured on an analog scale (±0.05 mg) after 120 days. The weight of the samples at 0% relative humidity was obtained by oven drying the samples at 105°C to a constant weight.

Vycor glass desorption isotherms were determined using initially water saturated glass samples (approximately 3 g each) immersed for a 1-month period in various sodium chloride solutions prepared at 0, 250, 500, and 1000 mmol/L. Once the immersion period was over, Vycor glass samples were then placed in ten different sealed conditioning chambers above different salt-saturated solutions, which kept the relative humidity constant.

The equilibrium moisture content of the specimens was measured on an analog scale (±0.05 mg) after a 14-day period. This short period of time was sufficient to get the equilibrium mass.

Results and Discussion

Water Vapor Transmission Tests (Test Series I)

Water vapor transmission test results are shown in Fig. 5. This figure presents the cumulative mass of moisture absorbed by the silica gel placed above the specimen as a function of time. Test results are presented for a unit surface area of mortar slabs. As can be seen, for both mortar mixtures (i.e., w/c = 0.40 and 0.60), the water vapor transmission test results are significantly affected by both the initial and the boundary (exposure) conditions.

For both water/cement ratios, the boundary conditions and the initial water content appear to have a marked influence on the moisture flux. The moisture flux emitted from each sample has been calculated on the basis of the test results presented in Fig. 5. Typical results for 0.60 w/c mortar samples are shown in Fig. 6.

As can be seen, the highest flux is always obtained for the initially dry mortar samples placed in direct contact with free water (see the experimental setup shown in Fig. 1[A]). For these specimens, the moisture flux measured during the first days of testing is quite important (nearly 500 and 800 g/d/m² for the 0.40 and 0.60 w/c ratio mortar samples, respectively), but decreases progressively to reach a steady-state regime after nearly 30 days of testing. At steady-state, the moisture flux values range from 60.4 g/d/m² for the 0.40 w/c ratio mixture to 139.0 g/d/m² for the 0.60 w/c mortar (see Table VI). This strong influence of water/cement ratio on the transport of moisture is in good agreement with the data reported by Brewer[17] for concrete samples.

For the initially dry samples placed in contact with saturated water vapor (see the experimental setup shown in Fig. 1[B]), the situation is quite different. Indeed, no significant amount of moisture was measured during the first 5–10 days of testing. However, after this period, a measurable amount of moisture began to be absorbed by the silica gel. After roughly 20 days of testing, the steady-state regime was established and moisture flux values of 13.0 and 31.9 g/d/m² were measured for the 0.40 and 0.60 w/c mortar samples, respectively (see Table VI).

Test results obtained for the initially saturated samples indicate that the moisture flux measured during the first few days of testing was unaffected

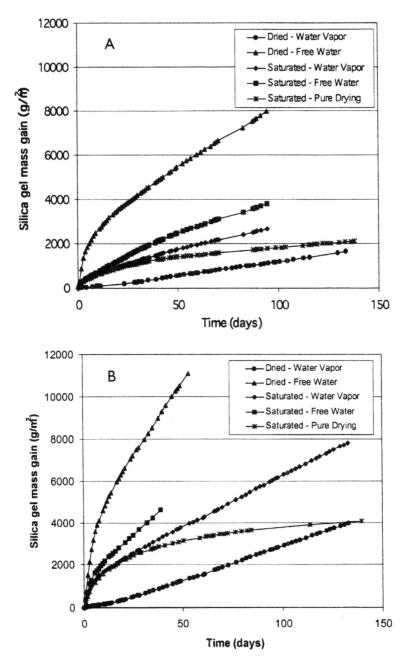

Figure 5. Water vapor transmission test results: (A) w/c = 0.40; (B) w/c = 0.60.

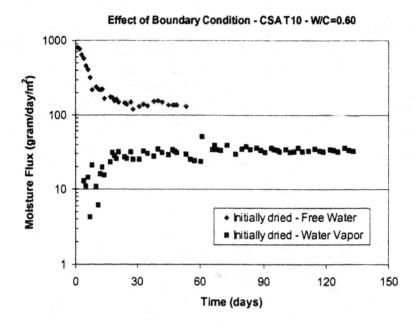

Effect of Boundary Condition - CSA T10 - W/C=0.60

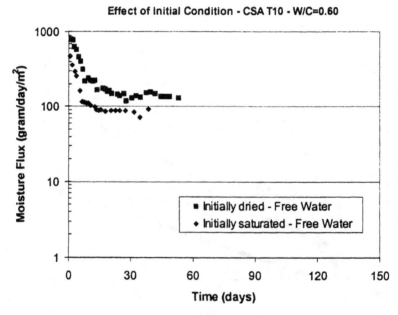

Effect of Initial Condition - CSA T10 - W/C=0.60

Figure 6. Effect of boundary and initial conditions on moisture flux (w/c = 0.60).

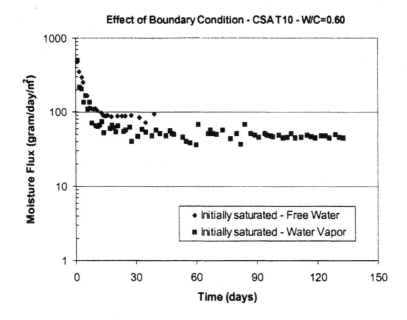

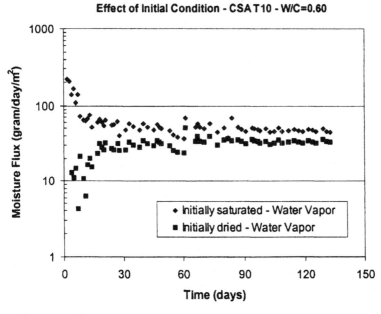

Figure 6, continued.

Table VI. Steady-state moisture flux measured on mortar slab (g/day/m²)

Mortar mixture w/c	Initial condition	Boundary condition	
		Liquid water	Saturated water vapor
0.40	Dried	60.4	13.0
0.40	Saturated	32.0	21.6
0.60	Dried	139.0	31.9
0.60	Saturated	86.8	49.4

by the boundary conditions. However, as testing continued, the moisture flux for the samples placed in a saturated vapor environment was found to be systematically lower than that measured for the specimens in contact with liquid water (see Table VI). This situation is likely related to the fact that the surface of mortar exposed to the saturated water vapor has not reached full saturation. This contributes to reducing the water content gradient across the sample (see Fig. 7). Accordingly, the driving force for the transport of moisture is decreased. The reduction in the water content of the mortar at the vicinity of the surface exposed to the water vapor is probably linked to the inability of the free water surface (located a few millimeters above the bottom surface of test sample) to maintain the air layer right next to the sample near saturation.

Results given in Table VI also show the marked influence of the initial moisture state of the material (i.e., initially dried or saturated) on the flux of moisture. As can be seen, the steady-state flux of the initially dry samples is roughly 1.5–2.0 times that measured for the saturated samples. Such a difference clearly shows the significant effect of the initial degree of saturation of the samples. The large difference between the two series of data is most probably link to the detrimental influence of the relatively harsh oven drying treatment on the microstructure of the mortar samples. Drying has been shown to coarsen the pore structure of hydrated cement systems.[18,19] These observations are in good agreement with the conclusions of Hazrati,[12] who observed that oven drying can significantly increase the kinetics of moisture transport in hydrated cement systems.

As part of this work, attempts were made to determine which percentage of the total flux being emitted from the sample was linked to the transmission of moisture through the material. Uniaxial drying test results (carried out on initially saturated samples) were compared to water vapor transmission data obtained for samples that had been saturated prior to testing. A

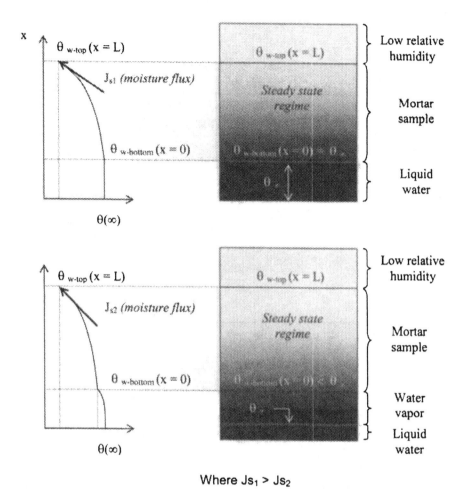

Where $Js_1 > Js_2$

Figure 7. Influence of boundary conditions on the steady-state moisture flux.

typical comparison of the two series of results obtained for the 0.40 w/c mortar samples is shown in Fig. 8. This figure indicates that, during the first days of testing, the flux of moisture for both test conditions cannot be distinguished. At this point, the total flux appears to be dominated by drying and the contribution of moisture migrating through the sample is not yet significant. However, as testing continued, a clear difference between the series of tests emerges. For instance, after 90 days of testing, the two fluxes differ by more than 60%. As can be seen, the data clearly emphasize the

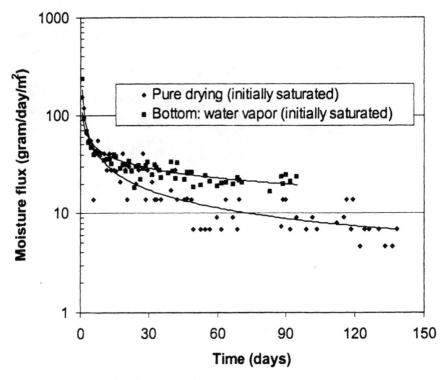

Figure 8. Contribution of water transmission through the sample to the total moisture flux emitted from the material (w/c = 0.40).

significant contribution of moisture transmission (i.e., moisture migrating through the sample) on the total flux of water determined during the test.

Water vapor transmission tests carried out in this study finally illustrate the marked influence of w/c on moisture transport. A reduction of w/c ratio from 0.60 to 0.40 tends to reduce the measured moisture flux by nearly 2–3 times. This dramatic influence of w/c is observed for all test conditions. Such results are in good agreement with the conclusions drawn by Brewer.[17]

The results of Test Series 1 have practical consequences for designers, concrete producers, and installers. The fact that significant quantities of moisture could be transmitted through the samples emphasizes the importance of considering the exposure conditions at the design stage of a structure. Although much thicker than the samples tested as part of this study, slabs on grade in direct contact with a source of moisture are likely to transmit excessive amounts of moisture (even after long periods of drying). This should be particularly the case for elements made of porous (high w/c)

Table VII. Water contents of the mortar samples (experimental data)

Mixture*	Initial water content (θ_0) (kg/m^3)	Capillary water content (θ_{cap}) (kg/m^3)	Vacuum water content (θ_{vac}) (kg/m^3)	Initial sample saturation (%)	Reference values[†]
Absorption tests					
T10 - 0.40 - 33% RH	67	171	182	0.36	0.38
T10 - 0.40 - 59% RH	91	162	182	0.50	0.52
T10 - 0.40 - 85% RH	133	164	177	0.75	0.70
T10 - 0.60 - 59% RH	75	201	217	0.35	0.33
T10 - 0.60 - 85% RH	107	187	211	0.50	0.50
Adsorption tests					
T10 - 0.40 - 33% RH	68		184	0.34	0.38
T10 - 0.40 - 59% RH	95		183	0.47	0.52
T10 - 0.40 - 85% RH	144		184	0.78	0.70

*RH: Relative humidity at which the sample has been exposed during the conditioning treatment.
[†]Reference value derived from desorption isotherms.

concrete. In that respect, a reduction of the w/c of the material presents numerous advantages. In addition to decreasing the time required to dry the material,[20] it also limits the transmission of moisture through the material.[17]

Absorption and Adsorption Test Results (Test Series 2)

As previously emphasized, Test Series 2 was performed to investigate the influence of the initial water content of the material on moisture uptake by capillary suction and vapor adsorption. This aspect is rarely discussed in papers devoted to the mechanisms of water transport in hydrated cement systems. The conditioning of samples (even relatively small ones like those used in this study) at different relative humidity levels is a time-consuming process. In addition, it is relatively difficult to verify if the material has reached equilibrium.

In this investigation, the initial degree of saturation and initial water content (θ_0) of all samples were determined prior to testing. In addition, at the end of the test, the capillary (θ_{cap}) and vacuum (θ_{vac}) water contents were also systematically measured. Results are summarized in Table VII. The capillary water content (θ_{cap}) is defined as the water content measured during the first stage of capillary suction tests (see Table VII).

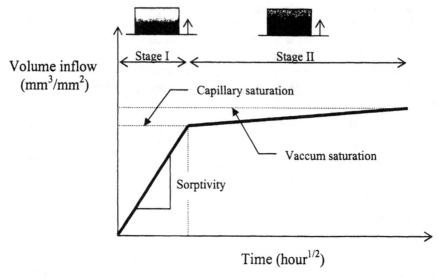

Figure 9. Typical absorption curve obtained in a free water uptake test (after Hazrati[12]).

In order to verify that the conditioning period had been sufficiently long to allow the samples to reach equilibrium, the initial sample saturation of each specimen was compared to a reference value derived from desorption isotherms obtained on companion samples. These reference values are also given in Table VII. As can be seen, both series of data are quite similar, suggesting that the test samples had indeed reached equilibrium.

Typical capillary absorption results are illustrated in Fig. 9. As can be seen, the uptake of water during such an experiment usually happens in two (more or less) distinct stages. The first one corresponds to the filling of all empty gel and capillary pores by water.[12] The second stage has been associated with the gradual filling of air-entrained bubbles and larger air voids resulting from an incomplete consolidation of the material.[21-23] As shown in Fig. 9, the filling of capillary and gel pore spaces is usually completed within a few hours, while the filling of larger voids is a much slower process that may take a few weeks.

Figures 10 and 11 present the experimental liquid water absorption and water vapor adsorption test results performed on the 0.40 and 0.60 w/c samples. In these figures, the last experimental point on each curve corresponds to the water content of the material after vacuum saturation. As it can be seen in both figures, the liquid water absorption curves all have sim-

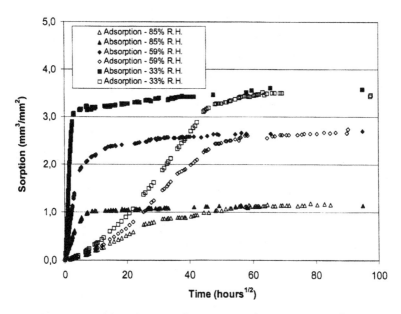

Figure 10. Experimental liquid water absorption and water vapor adsorption tests results carried out on 0.40 w/c mortar samples.

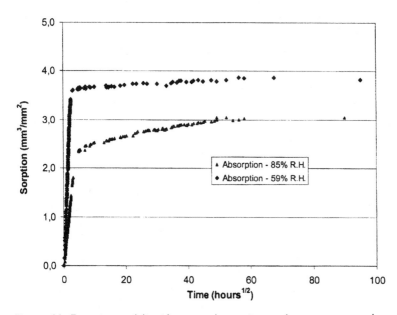

Figure 11. Experimental liquid water absorption and water vapor adsorption tests results carried out on 0.60 w/c mortar samples.

ilar shapes. In all cases but one, the two stages of water absorption are easily distinguished and well-defined nick points can be observed. The absorption of water during the initial stage appears to follow a square root of time law. Long-term water absorption (second stage) also seems to follow the same trend but the kinetics of water uptake during this second stage are much slower. Similar behaviors have been reported by Hazrati[12] and Martys and Ferraris.[1]

Test results shown in Fig. 10 also indicate that water vapor adsorption curves all have the same shape. However, in this case, the nick point is not as well defined as for absorption curves. Furthermore, test results indicated that the water vapor adsorption curves (obtained during the first stage) are not well described by a square root of time law.

Results presented in Figs. 10 and 11 clearly show the influence of boundary conditions on moisture transport in cement-based materials. Indeed, there is a marked difference between the kinetics of water absorption and that of vapor adsorption. For instance, during the absorption experiment, the filling of the initially empty gel and capillary pore spaces for the 0.40 w/c mortar equilibrated at 33% RH takes nearly 10 h to be completed. The time required to get the same material moisture content during water vapor adsorption test is increased to roughly 2000 h (85 days).

The difference between the two series of curves is linked to the basic mechanisms that control the uptake of water in each case.[22] During the absorption experiment, water is sucked into the pore structure of the material by strong capillary forces. Numerous test results suggest that the capillary forces act immediately when the surface of porous materials is placed in contact with liquid water.[12,24,25] During water vapor adsorption tests, capillary forces have (at least initially) no influence on the transport of water vapor molecules that penetrate the finely divided porous material by gaseous diffusion. Once within the material, water vapor molecules are then progressively adsorbed on the material pore walls. After a while, when the adsorbed water layers reach a critical thickness, capillary (or liquid water) condensation takes place, and meniscuses are formed at the narrow connections between pores (see Fig. 12). As pointed out by Xi et al.,[26] at this point, water molecules condense at one end of the pore neck, while at the other end they evaporate. The fact that vapor penetrates by diffusion explains why one cannot distinguish a nick point in the adsorption curves.

Despite the marked difference between the two series of curves, it is interesting to note that both mechanisms contribute ultimately to bring the samples to the same water content. This signifies that, at the end of the test-

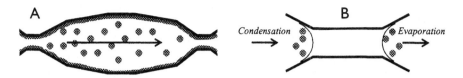

Figure 12. (A) Molecular diffusion and water vapor adsorption and (B) evaporation and condensation mechanism at pore neck.

ing period, all type of pores (i.e., gel and capillary pores, air bubbles, and compaction voids) are filled with the same amount of moisture, whether the sample has been subjected to water absorption or vapor adsorption.

As could be expected, test results also demonstrate that the kinetics of water absorption are particularly sensitive to the initial water content of the material. This effect can be best evaluated by calculating the slopes of the curves appearing in Figs. 10 and 11. This slope, often called the sorptivity of the material, is a measure of the velocity at which water penetrates the sample.[22,24] Results indicate that the sorptivity of all mortar samples is greatly reduced as the initial relative humidity inside the samples is increased (see Fig. 13). The values shown in Fig. 13 are close to those obtained by Hazrati[12] using mortar samples prepared with a white portland cement. As can be seen in Fig. 13, the sorptivity calculated for the 0.40 w/c mixture varies from 0.025 mm/min$^{1/2}$ (for the sample with a high degree of saturation) up to 0.130 mm/min$^{1/2}$ (for the sample tested at a low initial degree of saturation). These results indicate that the sorptivity of the material can be increased by more than five times when relative humidity within mortar samples is reduced from 85 to 33%. For mortar samples prepared at 0.60 w/c, sorptivity values vary from 0.074 mm/min$^{1/2}$ up to 0.150 mm/min$^{1/2}$.

Results presented in Fig. 13 also show the effect of w/c on the sorptivity of mortar samples. In this investigation, sorptivity was typically reduced by nearly 2.5–3.0 times when w/c was reduced from 0.60 to 0.40.

Significance of Test Results on Boundary Condition Modeling

The test results shown in Figs. 10 and 11 have important implications for the treatment of boundary conditions by numerical models aiming at describing the transport of moisture in hydrated cement systems. During a water absorption experiment, it appears that the capillary forces pulling liquid water inside the porous structure of the material act almost instantaneously when the surface of the material is placed in contact with liquid

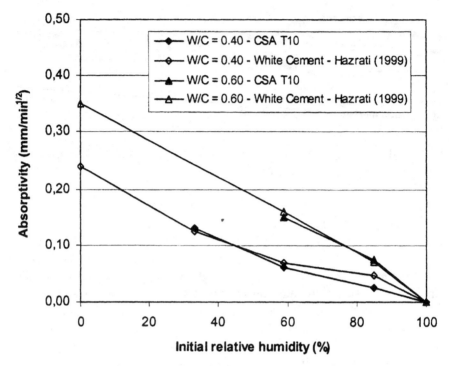

Figure 13. Influence of initial relative humidity on the sorptivity of mortars.

water. In this case, it is believed that moisture uptake during water absorption experiments should be modeled using the following boundary condition (Dirichlet's boundary condition):

$$H_\infty = H\ (x = 0, t) \tag{1a}$$

or

$$\theta_{w-\infty} = \theta_w\ (x = 0, t) \tag{1b}$$

where $H\ (x = 0, t)$ is the relative humidity at the material surface (%); H_∞ is the relative humidity of the air at the vicinity of the surface (%); $\theta_w\ (x = 0, t)$ is the moisture content at the material surface (m³/m³); and $\theta_{w-\infty}$ is the equivalent material moisture content (m³/m³).

Equation 1a is valid for models describing water transport in terms of variation of relative humidity. If the material water content (θ_w) is used as state variable (instead of relative humidity), H_∞ is then simply replaced by

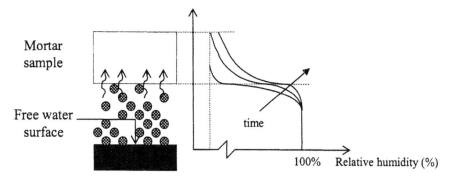

Figure 14. Water vapor molecule content at the vicinity of the porous material surface.

an equivalent material water content ($\theta_{w-\infty}$) given by the water vapor sorption isotherm of the material (see Eq. 1b).

The approach to be used for the treatment of water vapor adsorption experiments is slightly different. As previously explained, vapor molecules initially penetrate the porous material by gaseous diffusion. Accordingly, the water content of the material at the vicinity of the surface is reduced. The free water located a few centimeters below the surface of the tested sample will then try to reestablish the system equilibrium by releasing numerous water vapor molecules into the air. The latter phenomena (i.e., penetration of water vapor molecules in the material and the subsequent emission of water vapor molecules from the free liquid surface) is likely to create relative humidity gradient between the free water surface and the material surface (see Fig. 14). However, as the water vapor adsorption experiment progresses, the moisture content inside the material will increase and the water vapor diffusion kinetics will be slowed down. The reduction of the water vapor diffusion kinetics will then progressively diminish the capability of the material to reduce the water vapor content (or relative humidity) in the air at the vicinity of the sample surface. Accordingly, it is then suggested to model the water vapor adsorption in cementitious materials using the following type of boundary condition (Cauchy's boundary condition):

$$J_s = -D_m (H) (\partial H / \partial x) = \beta [H_\infty - H (x = 0, t)] \tag{2}$$

or

$$J_s = -D_m (\theta_w) (\partial \theta_w / \partial x) = \beta [\theta_{w\infty} - \theta_w (x = 0, t)] \tag{3}$$

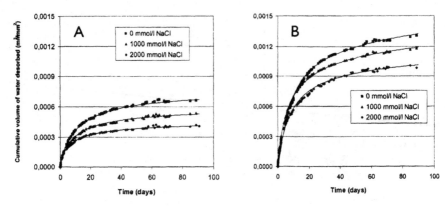

Figure 15. Water desorption during the drying experiments at 75% RH: (A) mortar samples with w/c = 0.40; (B) mortar samples with w/c = 0.60.

where J_s is the moisture flux at the material surface (kg/m²s); β is the exchange coefficient at the vicinity of the material surface (m/s); $D(H)$ is the moisture diffusivity at a given relative humidity (m²/s); and $D_m(\theta_w)$ is the moisture diffusivity at a given moisture content (m²/s).

Such an approach has been proposed in the past by various authors dealing with drying experiments (for instance, Mensi et al.[27]). The main difficulty of this treatment of the boundary conditions resides in the determination of the exchange coefficient β. Information on the influence of environmental conditions (i.e. wind velocity, temperature, etc.) on the value of β remains scarce. If attempts to investigate the effect of exchange conditions at the surface of the material on the transport of moisture have led researchers to propose guidelines for testing,[22,28,29] much remains to be done on the subject.

It should be emphasized that this approach may not be valid in cases where concrete is in contact with another porous medium. In that case, the use of a Dirichlet condition is probably still appropriate.

Drying of Mortar Samples Initially Immersed in Sodium Chloride Solutions (Test Series 3)

Experimental results obtained in this third test series are presented in Fig. 15. This figure shows the mass loss of the mortar samples upon drying. As previously described, all samples had been immersed in chloride solutions prior to testing. During the drying experiments, specimens were placed in a sealed chamber maintained at 75% RH at 23°C for a 3-month period. For both water/cement ratios (i.e., 0.40 and 0.60), test results indicate that the immersion in a sodium chloride solution has a significant influence on the drying kinetics of the samples. Indeed, an increase in the sodium chloride concentration reduces the rate of drying. For instance, after 3 months of exposure at 75% RH, the total amount of water lost from the 0.40 w/c sample initially immersed in a 2000 mmol/L NaCl solution is nearly 0.00041 mm³/mm² as compared to 0.00066 mm³/mm² for the reference sample (initially immersed in deionized water). As can be seen, the presence of sodium chloride in the material pore solution has reduced the loss of mass by 35–40%.

The same phenomenon is observed for the 0.60 w/c samples. The influence of the immersion in a sodium chloride solution, although significant, is less marked than what could be seen for the 0.40 w/c mixtures. The reduction in mass loss associated with the presence of high concentrations 2000 mmol/L in sodium chloride is limited to 30% (see Fig. 15). It should also be emphasized that, during the first 5–10 days of testing, sodium chloride had little, if any, effect on the mass loss curves. It is, however, difficult to establish if this observation is related to the mechanism of drying itself or simply to a limitation of the experimental technique used in this project.

As explained by Litvan,[30] the effect of sodium chloride on the kinetics of drying can be explained, at least in part, by the effect of sodium chloride on the thermodynamic properties of the solution. Sodium and chloride ions dissolved in solution contribute to modify the equilibrium pressure of the system. This effect tends to get more important as the drying process continues and gradually increases the ion concentration in the pore solution.

In addition, the presence of sodium and chloride ions also modifies the surface tension of the pore solution as described by the following equation[31]:

$$\sigma_{l/g-NaCl} = \sigma_{l/g-pure\ water} + 0.380\ [NaCl] \tag{4}$$

where $\sigma_{l/g-NaCl}$ is the surface tension of water in presence of NaCl (mN/m); $\sigma_{l/g-pure\ water}$ is the surface tension of pure water (mN/m); and [NaCl] is the sodium chloride concentration in water (mass%).

From Eq. 4, it can be seen that surface tension increases proportionally with the sodium chloride concentration of the pore solution. According to Laplace's equation, the latter phenomenon increases the capillary depression of water in the pore structure of the material:

$$\Delta P_c = P_l - P_g = (2\sigma_{l/g}) / r_p \tag{5}$$

where: ΔP_c is the capillary depression inside the material's pore structure (Pa); P_l is the pressure of the liquid phase inside the material's pore structure (Pa); P_g is the atmospheric pressure (Pa); and r_p is the pore radius (m).

According to Laplace's equation, when the surface tension of water is increased (due to the presence of sodium chloride in the pore solution), liquid water is then more tightly "trapped" within the porosity of the material. The main consequence of this phenomenon is obviously a reduction of the flux of water upon drying.

From a modeling point of view, it can be argued that sodium chloride in low and mid-range concentrations (~0–500 mmol/L) does not significantly affect the kinetics of drying of the material subjected to 75% RH. However, as the concentration in sodium and chloride ions is increased, their influence on drying is increasingly significant. In practice, many concrete structures are in contact with low and mid-range chloride concentrations, and the influence of these ions on drying can then reasonably be neglected in predictive moisture transport models. At high chloride concentrations (e.g., >500 mmol/L), it is believed that the effect of sodium chloride content should be considered. It should also be mentioned that the effect of sodium chloride at low and mid-range chloride concentrations may be significant for structures subjected to drying and wetting cycles. Under certain conditions, these cycles can favor a buildup in chlorides near the exposed surface. The influence of such a local increase in concentration should be evaluated on a case-by-case basis.

It should finally be emphasized that during this investigation, the influence of sodium chloride on water absorption was not investigated. This decision was made after the experimental results obtained by Perrin and Bonnet.[32] Indeed, according to these researchers, chloride salt concentration does not seem to have a great influence on liquid water absorption experiments, at least for 0.38 w/c mortar samples initially dried at 50°C.

Effect of Sodium Chloride on Desorption Isotherms (Test Series 4)

This last test series was carried out to establish if the equilibrium water content in cement-based materials (usually described by water sorption isotherms) is influenced by the presence of sodium chloride. Accordingly, water desorption isotherms were determined using both porous Vycor glass and neat cement pastes (prepared at 0.60 w/c using two portland cements). Porous Vycor glass was used to limit the influence of chemical reactions on the test results. Made of silicate, the Vycor glass samples are less likely to

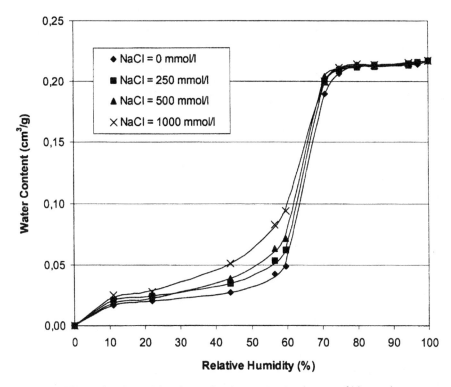

Figure 16. Effect of sodium chloride on the desorption isotherms of Vycor glass.

chemically interact with chloride ions. The other advantage of using Vycor glass samples resides in the fact that the relatively high porosity of the material allows reaching equilibrium with the external environment in less than 14 days, as compared to nearly 120 days for the powdered neat cement paste samples.

Vycor Glass Desorption Isotherms

Vycor glass samples were immersed in sodium chloride solutions prepared at 0, 250, 500, and 1000 mmol/L. The equilibrium water content was measured after 14 days of drying at a given relative humidity. Preliminary test results indicated that a 7-day period was sufficient to stabilize the mass loss of initially saturated Vycor glass samples used in this investigation.

Desorption isotherms obtained from drying experiments carried out in this investigation are presented in Fig. 16. All results were calculated in

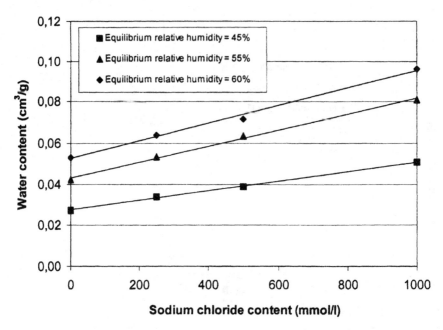

Figure 17. Effect of sodium chloride on the water content of Vycor glass.

terms of the relative pressure of pure water. As discussed by Litvan,[30] from a thermodynamic standpoint it would probably make more sense to express these results in terms of the partial pressure of the ionic solution found in the material's pores. However, such an exercise requires a good knowledge of the chemical composition of the pore solution, which is often hardly possible in practice.

Figure 16 clearly shows the influence of the sodium chloride concentration on the shape of desorption isotherm curves. As can be seen, the presence of sodium chloride modifies the water content of the material in the mid-range portion of the isotherm (25–75% RH) but has relatively little effect at high relative humidity (75–100% RH). Similar results were reported by Litvan.[30]

This particular behavior can partly be attributed to the fact that pores in the Vycor glass do not cover a wide range of sizes, their radii being mainly concentrated in the 40–50 Å range.[33,34] However, as explained by Litvan,[30] distortion is caused by the changing concentration of ions in solution during the course of the desorption process. As the water content of the material is decreased, the ionic strength of the solution gradually increases, which contributes to modify the equilibrium pressure of the pore solution. As previously discussed, the presence of ions also modifies the surface tension of the solution. A thorough discussion of these effects is beyond the scope of this paper. The reader is invited to refer to the very comprehensive analysis of Litvan[30] published many years ago.

In the mid-range relative humidity, the effect of sodium chloride concentration on material water content appears to be linear in nature (see Fig. 17). For instance, when the relative humidity is around 60%, the material water content can vary from 0.0530 cm^3/g (NaCl = 0 mmol/L) to 0.0960 cm^3/g (NaCl = 1000 mmol/L). When the sodium chloride concentration inside the material pore solution is increased from 0 to 2000 mmol/L, water content values are increased by more than 80%.

Cement Paste Desorption Isotherms

Based on test results obtained using Vycor glass samples, it was decided to run desorption experiments on neat cement pastes. Well-hydrated powdered cement paste samples (prepared at 0.60 w/c using CSA Type 10 and 50 portland cements) were immersed in sodium chloride solutions prepared at 0, 1000, 2000, and 4000 mmol/L. The equilibrium water content was measured after nearly 120 days of drying. Preliminary test results had indicated that this delay was sufficient to stabilize the mass loss of the neat paste samples.

Desorption isotherms are presented in Figures 18 and 19. In the figures, all results are expressed in terms of the partial pressure of pure water. These figures clearly show the marked influence of sodium chloride concentration on the general shape of desorption isotherm curves for both cement pastes tested. It is, however, important to emphasize that the effect of sodium chloride appears to be solely limited to relative humidity levels higher than 45–50%. The latter observation is particularly interesting. Indeed, as pointed out by Mills,[35] when the relative humidity inside a cementitious material is lower than 45%, no meniscus can be sustained in the material's porous network because the surface tension exceeds the tensile strength of water. Consequently, test results obtained here seem to indicate that the effect of sodium chloride salt is solely limited to capillary water, which can be found in cement-based materials exposed to high and mid-range relative humidity levels.

When the relative humidity reaches the 75–100% range, the desorption isotherm presented in Fig. 18 indicates that the presence of sodium chloride

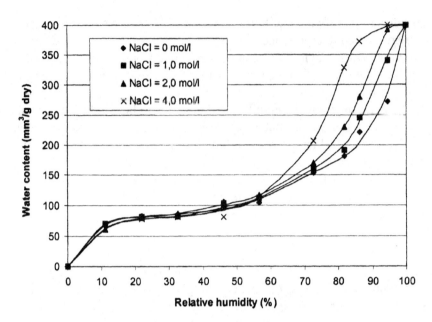

Figure 18. Effect of sodium chloride on the desorption isotherms of CSA T10 neat cement paste samples (w/c = 0.60).

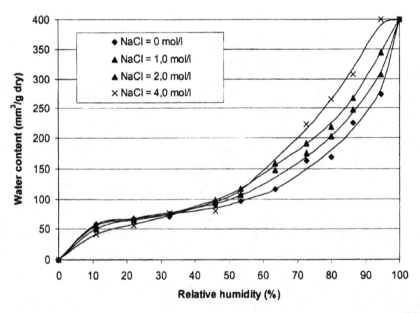

Figure 19. Effect of sodium chloride on the desorption isotherms of neat CSA T50 neat cement paste samples (w/c = 0.60).

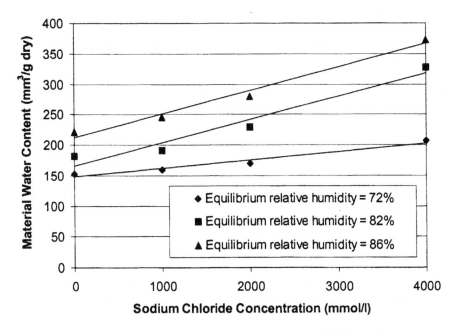

Figure 20. Effect of sodium chloride on the water content of CSA Type 10 neat cement paste.

can significantly increase the water content of the material. For instance, when the relative humidity is around 80%, the material water content varies from 180 cm^3/g (NaCl = 0 mmol/L) to 230 cm^3/g (NaCl = 2000 mmol/L). The material water content is therefore increased by 28% when the sodium chloride concentration increases from 0 to 2000 mmol/L.

These results are interesting from a modeling point of view since many structures are often exposed to drying at such relative humidity levels. They also indicate that the presence of sodium chloride salt tends to increase the equilibrium water content, which reduces the material water content gradient during a drying experiment. This reduction of material water content gradient obviously reduces the flux of moisture inside the material. Such a phenomenon was observed during the drying experiments discussed in the previous section. Finally, as for the Vycor glass experiments, the influence of sodium chloride salt on material water content appears to be linear in nature in the mid-range relative humidity, at least when results are expressed in terms of the partial pressure of pure water (see Fig. 20). This apparent linear behavior makes this phenomenon relatively easy to take into account in moisture transport models.

Conclusion

The results of an investigation carried out to quantify the influence of initial and boundary conditions on moisture transport in cement-based materials have been presented in this chapter.

Water vapor transmission test results (Test Series 1) have first shown that boundary and initial conditions appear to have a significant influence on the moisture flux. As part of the investigation, the highest flux was always obtained on initially dry samples placed in direct contact with a free liquid water surface. The lowest moisture flux was obtained on initially dry samples placed in contact with saturated water vapor (100% RH). For the initially dry samples (w/c = 0.40 and 0.60), the flux of moisture measured on samples placed in direct contact with a free liquid water surface was nearly 4.5 times higher than the one measured on samples placed in contact with a saturated water vapor. For initially saturated mortar samples the difference is still important but slightly lower (around 1.5 times). Finally, test results obtained in this first test series have clearly shown the significant contribution of moisture transmission through the sample on the total amount of water emitted from the surface.

Liquid water absorption and water vapor adsorption tests (Test Series 2) were carried out on mortar samples (w/c = 0.40 and 0.60). Test results have clearly shown that there is a marked difference between liquid water absorption and water vapor adsorption, whatever the tested samples. Typically, for the absorption tests, the filling of initially empty gel and capillary pores takes only a few hours to be completed, as compared to several days for the water vapor adsorption tests. However, both phenomena ultimately brought the samples to the same final degree of saturation.

Drying (at 75% RH) of mortar samples initially immersed in salt solutions (Test Series 3) has indicated that the presence sodium chloride has a significant influence on the drying kinetics of tested samples. As the sodium chloride concentration in the material is increased, the kinetics of drying are reduced. According to the data obtained in this investigation, the presence of sodium chloride in the material pore solution has reduced the loss of mass by nearly 30–40%, depending on the w/c ratio of the tested materials. It is believed that the effect of sodium chloride on the rate of mass loss upon drying can be, at least in part, attributed to a modification of the liquid water surface tension in presence of sodium chloride.

The last series of tests (Test Series 4) has established that the equilibrium water content in cement-based materials (given by water desorption isotherms) is influenced by the presence of sodium chloride in the material's pore solution. However, desorption tests carried out on crushed cement paste samples have indicated that this effect is solely limited to mid-range and high relative humidity levels (i.e., RH higher than 45–55%).

References

1. N. S. Martys and C. F. Ferraris, "Capillary Transport in Mortars and Concrete," *Cem. Concr. Res.*, **27** [5] 747–760 (1997).
2. A. Bentur, S. Diamond, and N. Berke, *Steel Corrosion in Concrete: Fundamentals and Civil Engineering Practice.* E&FN Spon, London, 1999.
3. J. Skalny, J. Marchand, and I. Odler, *Sulfate Attack on Concrete.* Modern Concrete Technology 10. E&FN Spon, London, 2002.
4. A. Goudie and H. Viles, *Salt Weathering Hazards.* Wiley, New York, 1997.
5. C. Rodrigues-Navarro, E. Doehne, and H. Sebastian, "How Does Sodium Sulfate Crystallize? Implication for the Decay and Testing of Building Materials," *Cem. Concr. Res.*, **30**, 1527–1534 (2000).
6. N. Tsui, R. J. Flatt, and G. W. Scherer, "Crystallization Damage by Sodium Sulfate," *J. Cultural Heritage*, **4**, 109–115 (2003).
7. G. W. Scherer, "Stress from Crystallization of Salt," *Cem. Concr. Res.* (in press).
8. B. A. Surprenant and W. R. Malish, "Are Your Slabs Dry Enough for Floor Covering?"; pp. 42–47 in *Moisture Problems in Concrete Floors: Analysis and Prevention.* American Concrete Institute, 1998.
9. R. W. Day, "Moisture Migration through Concrete Floor Slabs," *J. Perform. Constructed Facilities*, **6** [1] 46–51 (1992).
10. S. L. Sarkar, "Understanding Floor Covering Failure Mechanisms," *Concr. Repair Bull.*, March/April 1999, pp. 6–11.
11. L.-O. Nilsson and J.-P. Ollivier, "Fundamentals of Transport Properties of Cement-Based Materials and General Methods to Study Transport Properties"; pp. 113–148 in *Engineering and Transport Properties of the ITZ in Cementitious Composites.* RILEM Report 20. 1999.
12. K. Hazrati, "Study of Water Absorption Mechanisms in Ordinary and High-Performance Mortars," Ph.D. thesis, Laval University, Canada, 1998.
13. J. Selih, A. Sousa. and W. Bremner, "Moisture Transport in Initially Fully Saturated Concrete During Drying," *Transport in Porous Media*, **24**, 81–106 (1996).
14. S. Jacobsen, B. Gérard, and J. Marchand, "Prediction of Short Time Drying for OPC and Silica Fume Concrete Frost/Salt Scaling Testing"; pp. 401–410 in *Durability of Building Materials and Components VII*, vol. 1. E&FN Spon, London, 1996.
15. D. Xin, D. G. Zollinger, and G. D. Allen, "An Approach to Determine Diffusivity in Hardening Concrete Based on Measured Humidity Profiles," *Adv. Cem. Based Mater.*, **2** [4] 138–144 (1995).
16. J. F. Daian, "Condensation and Isothermal Water Transfer in Cement Mortar, Part I: Pore Size Distribution, Equilibrium, Water Condensation and Inhibition," *Transport in Porous Media*, **3**, 563–589 (1988).

17. H. W. Brewer, "Moisture Migration: Concrete Slab-on-Ground Construction," *J. PCA Res. Devel. Lab.*, pp. 2–24 (1965).
18. D. H. Bager and E. J. Sellevold, "Ice Formation in Hardened Cement Paste, Part II: Drying and Resaturation of Room Temperature Cured Paste," *Cement Concr. Res.*, **16** [6] 835–844 (1986).
19. R. F. Feldman, "Effect of Pre-Drying on Rate of Water Replacement from Cement Paste by Propan-2-ol," *Il Cemento*, **3**, 193–201 (1988).
20. B. A. Surprenant and W. R. Malish, "Don't Puncture the Vapor Retarder"; pp. 46–48 in *Moisture Problems in Concrete Floors: Analysis and Prevention.* American Concrete Institute, 1998.
21. G. Fagerlund, "Predicting the Service Life of Concrete Exposed to Frost Action through a Modeling of the Water Absorption Process in the Air-Pore System"; pp. 503–537 in *The Modeling of Microstructure and Its Potential for Studying Transport Properties and Durability.* Edited by H. Jennings. Pp. 503–537. Kluwer, 1996.
22. G. Hedenblad, "Moisture Permeability of Mature Concrete, Cement Mortar and Cement Paste," Ph.D. thesis, Division of Building Materials, Lund Institute of Technology, 1993.
23. J. Punkki and E. J. Sellevold, "Capillary Suction in Concrete: Effects of Drying Procedure," *Nordic Concr. Res.*, **1**, 101–116 (1994).
24. C. Hall, "Barrier Performance of Concrete: A Review of Fluid Transport Theory," *Mater. Struct.*, **27**, 291–306 (1994).
25. L. Pel, "Moisture Transport in Porous Building Materials," Ph.D. thesis, Eindhoven University of Technology, Netherlands, 1995.
26. Y. Xi, Z. P. Bazant, and H. M. Jennings, "Moisture Diffusion in Cementitious Materials: Adsorption Isotherms," *Adv. Cem. Based Mater.*, **1** [6] 248–257 (1994).
27. R. Mensi, P. Acker, and A. Attolou, "Séchage du béton: Analyse et modélisation," *Mater. Struct.*, **21**, 3–21 (1988).
28. M. Bomberg, "Testing Water Vapor Transmission: Unresolved Issues"; pp. 157–167 in *Water Vapor Transmission through Building Materials and Systems: Mechanisms and Measurements.* ASTM STP 1039. ASTM, 1987.
29. G. H. Galbraith, R. C. McClean, and Zhi Tao, "Vapour Permeability: Suitability and Consistency of Current Test Procedures," *Building Serv. Eng. Res. Technol.*, **14** [2] 67–70 (1993).
30. G. G. Litvan, "Adsorption Systems at Temperatures Below the Freezing Point of the Adsorptive," *Adv. Colloid Interface Sci.*, **9**, 253–302 (1978).
31. Davide R. Lide, ed. *CRC Handbook of Chemistry and Physics.* CRC Press Inc., Boca Raton, Florida, 1999.
32. B. Perrin and S. Bonnet, "Experimental Results Concerning Combined Transport of Humidity and Chloride in Non-Steady-State," *Concr. Sci. Eng.*, **2** [6] 117–124 (2000).
33. P. Levitz, G. Ehret, S. Sinha, and J. Drake, "Porous Vycor Glass: The Microstructure as Probed by Electron Microscopy Direct Transfer, Small-Angle Scattering and Molecular Absorption," *J. Chem. Phys.*, **95** [8] 6151–6161 (1991).
34. G. G. Litvan, "Phase Transitions of Adsorbates, Part I: Specific Heat and Dimensional Changes of the Porous Glass–Water System," *Canadian J. Chem.*, **44**, 2617–2622 (1966).
35. R. H. Mills, Mass Transfer of Water Vapor through Concrete, *Cem. Concr. Res.*, **15**, 74–82 (1995).

Ionic Interactions in Cement-Based Materials: The Importance of Physical and Chemical Interactions in the Presence of Chloride or Sulfate Ions

Y. Maltais, J. Marchand, P. Henocq, T. Zhang, and J. Duchesne

Ionic interactions in cement-based materials were investigated using tricalcium silicate pastes and portland cement paste systems. Cementitious materials were immersed in saturated lime or alkaline solutions prepared at various sodium chloride or sodium sulfate concentrations. The amount of bound chloride or sulfate ions was determined by measuring the variation in chloride or sulfate concentrations of the test solution after 4 months of immersion. Immersion tests were conducted on cementitous materials hydrated for at least 2 years. Three types of cement (ASTM Type I, ASTM Type V, and a white cement) and two water/cement ratios (0.40 and 0.60) were used. Test results show that electrostatic interactions dominate the interaction of chloride ions with tricalcium silicate pastes, specific adsorption being negligible. For sulfate ions, test results suggest that physical binding is the sum of both electrostatic interaction and specific adsorption. As expected, both chemical and physical interactions were found to be involved in the binding of chloride ions to neat cement pastes. Test results also suggest that physical binding plays only a marginal role in the binding of sulfate ions, which is essentially controlled by the precipitation of new solid phases (chemical binding).

Introduction

Concrete degradation mechanisms generally involve the penetration of external ions (such as chloride, sulfate, and magnesium) into the material porosity and/or the dissolution of various hydrated and unhydrated phases. The prediction of the degradation of concrete over time warrants a good knowledge of the properties of the material, such as its ability to transport ions and moisture. The mechanisms by which ions interact with the solid phases of the material are also greatly important. Not only do they play a significant role in the development of deleterious chemical reactions, but they also affect the transport of ions through the material porous structure.

Over the past decades, it has been clearly established that ions can physically and chemically interact with the hydrated cement paste.[1,2] It is, for instance, well known that some ions (such as chlorides and sulfates) can react with the unhydrated and hydrated phases of cement to form new compounds.[3] It has also been clearly established that ions can undergo physical interaction with hydrated cement phases.[4-8] According to some recent reports, it seems that physical binding is the sum of surface complexation phenomena and electrostatic interactions arising from the presence of short-range electrical forces at the pore surface.

Reports discussing the relative contribution of each phenomenon (i.e., chemical binding, surface complexation, and electrostatic interaction) to the binding of ions with hydrated cement systems remain scarce. This situation is mainly due to the fact that most investigations where carried out on neat cement paste or mortar samples (see, for instance, Refs. 9 and 10). During these experiments, the binding of ions is often dominated by the precipitation (and dissolution) of solid phases. The need for such a distinction has, however, recently become important, since new service-life prediction models often address physical and chemical interaction mechanisms separately.[11] In order to investigate the problem, Wowra and Setzer,[6] Catinaud,[12] and Zibara[10] have recently used both tricalcium silicate pastes and portland cement pastes to discuss the relative importance of physical and chemical interactions with chloride ions.

This paper attempts to evaluate the relative importance of physical and chemical binding mechanisms of chloride and sulfate ions with hydrated cement systems by comparing experimental and numerical results obtained on well-hydrated tricalcium silicate mixtures and neat portland cement paste samples. This paper also provides information on the parameters influencing the interaction of ions with the cement paste.

Theoretical Consideration Pertaining to Ionic Interaction in Cement-Based Materials

Physical Binding Mechanisms

Over the past decade, the mechanisms of physical binding (also referred to as physical adsorption) have received relatively little attention from researchers involved in the field of cement science. As pointed out by Castellote et al.,[2] the difficulty of delineating physical interaction from chemical reactions has often forced researchers to study ionic binding mechanisms using a global approach without making any distinctions between the two phenomena. The problem is further complicated by the fact that physical binding of ions appear to be the sum of electrostatic inter-

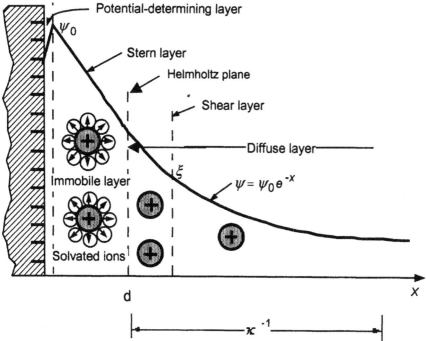

Figure 1. Formation of the double layer at the vicinity of the pore wall. (Adapted from Ref. 15.)

actions and surface complexation phenomena. These two mechanisms are briefly discussed in the following subsection.

Electrostatic Interaction

Experimental evidence suggests that the physical binding of ions with the cement paste hydrates arises mainly from short-range electrostatic forces resulting from the presence of electrical charges at the surface of the solid.[7,8,13,14] These charges impose a constraint on the distribution of ions at the vicinity of the solid/liquid interface in an attempt to restore electroneutrality. This reorganization of charges results in the formation of an electrical double layer (see Fig. 1). This is done by an accumulation of charges in a diffuse layer near the interface, thus creating a nonzero electrical potential in this area.

The potential measured at the interface between the bound ion layer and the pore solution is called the zeta potential. Reports indicate that the zeta potential of hydrated cement systems may vary from –35 mV up to +35 mV.[7,8,14] The evolution of the electrical potential in the double layer is given by the Poisson equation[16]:

$$\frac{\partial^2 \phi(x)}{\partial x^2} = -\frac{F}{\varepsilon} \sum_i z_i c_i (x = \infty) \exp\left(\frac{-z_i F}{RT} \phi(x)\right) \quad (1)$$

where $\phi(x)$ is the electrical potential, x is the distance from the surface, ε is permittivity of the medium, F is the Faraday constant, z_i is the ionic valence, $c_i(x = \infty)$ is the bulk solution concentration, R is the perfect gas constant, and T is the temperature.

Ions at the vicinity of the interface are tightly bound to the first layer of charges and they cannot participate, at least from a statistical standpoint, in the transport process.[15] These ions are then considered to be physically bound to cement hydrated phases. It is usually admitted that the region over which the behavior of ions is affected by the electrical forces extends from the surface of the pore to a distance approximately equal to the Debye-Huckel length (κ^{-1}) within the solution. The expression of the Debye-Huckel length is given by the following equation:

$$\kappa = \left(\frac{F^2}{\varepsilon RT} \sum_i z_i^2 c_i (x = \infty)\right)^{1/2} \quad (2)$$

Surface Complexation Phenomena

Investigations suggest that surface complexation may also participate in the physical binding of ions by hydrated cement systems. These surface complexation phenomena involve the exchange of one or numerous ionic species between the surface of the solid and the aqueous phase.[17] In the literature, this type of surface chemical reaction has also been referred to as specific ionic adsorption.[8] According to Wowra and Setzer[6] and Viallis-Terrisse,[8] there is strong evidence that calcium ions (Ca^{2+}) are being specifically adsorbed to the solid surface. These conclusions were derived from significant variations of the zeta potential measured on some C-S-H suspensions equilibrated at different CaO concentrations (see Fig. 2). According to Nachbaur et al.,[7] similar results were also observed for sulfate ions (see Fig. 2). It is important to underline that no such effect could be observed for other species such as sodium (Na^+), potassium (K^+), and even chloride (Cl^-).[6,7,13]

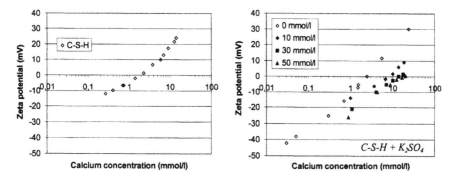

Figure 2. Typical zeta potential variation for different cementitious systems. (Reproduced from Refs. 7 and 8.)

As for the formation of new compounds, surface complexation phenomena can be modeled using stochiometric and equilibrium equations. For instance, the surface complexation of calcium ions can be described by the following equations[8]:

$$SiOH + Ca^{2+} \leftrightarrow SiOCa^+ + H^+ \qquad (3)$$

with

$$K = \frac{\{SiOCa^+\}\{H^+\}}{\{SiOH\}\{Ca^{2+}\}} \qquad (4)$$

The previous equations describe a ligand (SiO$^-$) exchange reaction at the C-S-H surface with an equilibrium constant K. As pointed out by Viallis-Terrisse,[8] surface reactions with the silanol (SiOH) groups can also be considered with similar equations.

Chemical Interactions

As emphasized by many researchers, the phase transformations that may occur in hydrated cement systems in contact with chloride or sulfate solutions are quite intricate. In fact, a number of chemical reactions can occur that may lead, for instance, to the formation of new products and the dissolution of phases already present in the material. It is now well known that chloride ions can react with aluminate phases to form new chloride-bearing phases such as the following[3]:

Table I. Solubility of some hydrated solid phases in cement-based materials[21,22]

Name	Chemical composition	Expression for equilibrium	K_{sp}
Portlandite	$Ca(OH)_2$	$K_{sp}=\{Ca\}\{OH\}^2$	5.2
Ettringite	$3CaO \cdot Al_2O_3 \cdot 3CaSO_4 \cdot 32H_2O$	$K_{sp}=\{Ca\}^6\{OH\}^4\{SO_4\}^3\{Al(OH)_4\}^2$	44.4
Friedel's salt	$3CaO \cdot Al_2O_3 \cdot CaCl_2 \cdot 10H_2O$	$K_{sp}=\{Ca\}^4\{OH\}^4\{Cl\}^2\{Al(OH)_4\}^2$	29.1
Hydrogarnet	$3CaO \cdot Al_2O_3 \cdot 6H_2O$	$K_{sp}=\{Ca\}^3\{OH\}^4\{Al(OH)_4\}^2$	23.2
Gypsum	$CaSO_4 \cdot 2H_2O$	$K_{sp}=\{Ca\}\{SO_4\}$	4.6
Mirabilite	$Na_2SO_4 \cdot 10H_2O$	$K_{sp}=\{Na\}^2\{SO_4\}$	1.2

- Calcium monochloroaluminate or Fridel's salt ($3CaO \cdot Al_2O_3 \cdot CaCl_2 \cdot 10H_2O$).
- Calcium trichloroaluminate ($3CaO \cdot Al_2O_3 \cdot 3CaCl_2 \cdot 10H_2O$).
- Chloroferrite hydrates ($3CaO \cdot Fe_2O_3 \cdot CaCl_2 \cdot 10H_2O$).

Numerous investigations[18-20] have also shown that sulfate ions can interact with aluminate phases to form new sulfate-bearing phases such as:

- Calcium sulfoaluminate or ettringite ($3CaO \cdot Al_2O_3 \cdot 3CaSO_4 \cdot 32H_2O$).
- Calcium sulfate or gypsum ($CaSO_4$).

Over the past few years, some numerical models have been developed to study the influence of various ions (such as chlorides and sulfates) on the chemical equilibrium of cementitious systems (see, for instance, Refs. 21 and 22). Those numerical models account for the solubility of each solid phase (see Table I), the electro-neutrality, and the chemical activity of the pore solution.

Test Program

The test program of the present investigation was divided in two distinct parts. The first part was devoted to the study of physical interactions on well-hydrated pure C_3S pastes prepared at a 0.50 water/binder ratio. Interaction mechanisms were investigated by measuring chloride and sulfate binding isotherms. Test variables included the composition of the immersion solution (sodium chloride solution or sodium sulfate solution) and the nature and ionic strength of the base solution (lime solution or alkaline solution). Binding isotherms were compared to numerical simulations results carried out using Poisson equation (see previous section).

The objective of the second test series was to investigate the relative importance of physical and chemical interaction phenomena in hydrated cement systems. Cement pastes used in this second test series were prepared at two different water/cement ratios. Test variables also included type of cement, nature of the immersion solution (sodium chloride solution or sodium sulfate solution), nature and ionic strength of the base solution (lime solution or alkaline solution), and initial water content of the material (initially dried or initially saturated).

Materials and Experimental Procedures

Materials and Mixture Characteristics

As previously pointed out, neat cement pastes used in this investigation were prepared at water/cement ratios (w/c) of 0.40 and 0.60. Cements pastes were prepared using a CSA Type 10 portland cement (ASTM Type I equivalent), a CSA Type 50 portland cement (low C_3A content; ASTM Type V equivalent), and a white portland cement (low C_4AF content). The chemical and mineralogical compositions of each cement are given in Table II. The tricalcium silicate pastes used in this investigation were supplied by the National Research Council of Canada (Institute of Research in Construction in Ottawa). These well-hydrated pastes were prepared at 0.50 water/binder ratio using pure C_3S phase (CaO and SiO_2 only).

All cement paste mixtures were prepared using deionized water. The mixtures were batched in a high-speed mixer placed under vacuum (at 10 mbar) to prevent, as much as possible, the formation of air voids during mixing. Mixtures were cast in plastic cylinders (diameter = 7.0 cm; height = 20 cm). The molds were sealed and rotated for the first 24 h in order to prevent any bleeding of the mixtures. At the end of this period, the cylinders were demolded and sealed with an adhesive aluminum foil for a 24-month period at room temperature. This period was selected to ensure well-cured cement systems.

Experimental Procedures

Tricalcium silicate and portland cement pastes were cut in thin slices of 2 mm. Except for the ionic binding tests carried out on saturated samples (see previous section), each slice was vacuum dried for a minimum period of 28 days and then kept at a relative humidity of 11% (using a LiCl solution) until testing. Samples (approximately 2.0 g each for C_3S pastes and 10 g each for cement pastes) were then immersed in lime or alkaline solutions containing different chloride or sulfate concentrations. The volume of test

Table II. Oxide and mineralogical compositions (%) and LOI of the three cements

	CSA Type 10	CSA Type 50	White
Oxides			
SiO_2	19.78	21.45	24.27
Al_2O_3	4.39	3.58	1.84
TiO_2	0.22	0.21	0.09
Fe_2O_3	3.00	4.38	0.30
CaO	62.04	63.93	68.83
SrO	0.26	0.07	0.12
MgO	2.84	1.81	0.64
Mn_2O_3	0.04	0.05	0.01
Na_2O	0.32	0.24	0.17
K_2O	0.91	0.70	0.03
SO_3	3.20	2.28	2.14
Compounds			
C_3S	59	62	77
C_2S	12	16	12
C_3A	7	2	4
C_4AF	9	13	1
LOI	2.41	0.86	1.18

solutions was fixed at 15 and 50 mL for C_3S paste samples and cement paste samples, respectively. Samples were left immersed in solution for four months, a period sufficient to reach chemical equilibrium between the solids and the immersion solution. At the end of the immersion period, the chloride or sulfate concentrations of the solutions were determined by means of potentiometric titration.

Tables III and IV present a summary of the experimental conditions investigated during the course of this work. Chloride and sulfate contents were selected to account for typical concentrations usually found in practice (for instance, the chloride concentration of seawater on the east coast of Canada is in the range 500–1000 mmol/L and sulfate content of soils in the western provinces is usually between 0 and 50 mmol/L). For interaction tests carried out on C_3S pastes, base solution concentrations were selected in order to show the effect (if any) of alkaline content on the binding mechanisms. For the interaction tests performed on cement pastes, base solution concentrations were determined using real pore solution alkaline content of

Table III. Experimental conditions for hydrated tricalcium silicate mixtures

Base solution	Aggressive solution	
	Nature of the solution	Concentration (mmol/L)
Ca(OH)$_2$ (saturated lime solution)	NaCl	10, 25, 50, 100, 250, 500
	Na$_2$SO$_4$	10, 25, 50
NaOH (320 mmol/L)	NaCl	10
	Na$_2$SO$_4$	10
NaOH (110 mmol/L) + KOH (325 mmol/L)	NaCl	10, 50, 250
	Na$_2$SO$_4$	10

each neat cement paste. The objective was to investigate binding mechanisms under experimental conditions as close as possible to the real alkaline content of the different hydrated cement systems.

Calculation of Bound Chloride and Sulfate Contents

The amount of bound chloride or sulfate ions was calculated according to the following equation[3]:

$$[X]_{bound} \frac{[X]_{initial} - [X]_{equilibrium}}{W} V \tag{5}$$

where $[X]_{bound}$ is bound chloride or sulfate content (mg/g sample); $[X]_{initial}$ is initial chloride or sulfate content (mg/L); $[X]_{equilibrium}$ is equilibrium chloride or sulfate content (mg/L); V is the volume of the immersion solution (L); and W is sample weight (g).

As mentioned by Delagrave[3] and Tang and Nilsson,[23] the amount of bound ionic species can also be expressed on a unit of cement gel basis using the following equation:

$$W_{gel} = \frac{\left(1 + W_n^0\right) f_c \alpha}{1 + W_n^0 f_c \alpha} W \tag{6}$$

where W_n^0 is the maximum chemically bound water content (this value usually varies from 0.22 to 0.25 mg/g sample), α is the degree of hydration of the cement, and f_c is the relative cement paste content of the mixture ($f_c = 1$ for neat cement pastes).

Table IV. Experimental conditions for hydrated cement paste mixtures

Material	Immersion solution	Sample preparation	Aggressive solution
CSA T10 - w/c = 0.40	Ca(OH)$_2$	Saturated	NaCl (1–1000 mmol/L)
		Dried	NaCl (1–1000 mmol/L)
	NaOH (195 mmol/L) + KOH (590 mmol/L)	Saturated	NaCl (1–1000 mmol/L)
		Dried	NaCl (1–1000 mmol/L)
CSA T10 - w/c = 0.60	Ca(OH)$_2$	Saturated	NaCl (1–1000 mmol/L)
		Dried	NaCl (1–1000 mmol/L)/ Na$_2$SO$_4$ (1–100 mmol/L)
	NaOH (110 mmol/L) + KOH (325 mmol/L)	Saturated	NaCl (1–1000 mmol/L)/ Na$_2$SO$_4$ (1–100 mmol/L)
		Dried	NaCl (1–1000 mmol/L)
CSA T50 - w/c = 0.40	Ca(OH)$_2$	Saturated	NaCl (1–1000 mmol/L)
	NaOH (170 mmol/L) + KOH (415 mmol/L)	Saturated	NaCl (1–1000 mmol/L)
CSA T50 - w/c = 0.60	Ca(OH)$_2$	Saturated	NaCl (1–1000 mmol/L)
	NaOH (110 mmol/L) + KOH (220 mmol/L)	Saturated	NaCl (1–1000 mmol/L)/ Na$_2$SO$_4$ (1–100 mmol/L)
White - w/c = 0.40	Ca(OH)$_2$	Saturated	NaCl (1–1000 mmol/L)
	NaOH (130 mmol/L) + KOH (30 mmol/L)	Saturated	NaCl (1–1000 mmol/L)

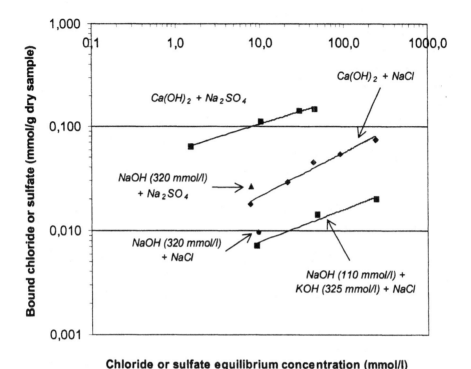

Figure 3. Chloride and sulfate binding isotherms obtained for hydrated C₃S pastes.

Test Results

Binding Isotherms: Hydrated Tricalcium Silicate Samples

Chloride and sulfate binding isotherms obtained on hydrated tricalcium silicate pastes are shown in Fig. 3. In this figure, the amount of chlorides or sulfates bound to the solids is given as a function of the equilibrium chloride or sulfate concentration measured in the immersion solution upon the completion of the binding experiments.

The test results presented in Fig. 3 clearly indicate that the amount of bound ions strongly depends on the equilibrium concentration of the immersion solution. Similar results are commonly reported for cement pastes.[24] The shape of the binding isotherms shown in Fig. 3 suggests that the ionic interaction mechanisms are nonlinear in nature and can be modeled using a classical Freundlich isotherm:

$$\log[X]_{bound} = \alpha \log[X]_{equilibrium} + \beta \qquad (7)$$

As pointed out by Mejlhede Jensen et al.,[25] the Freundlich isotherm equation is, unfortunately, semi-empirical, and the parameters α and β have no real physical meaning.

The interaction isotherms presented in Fig. 3 clearly show the influence of the chemical composition of the base solution (i.e., lime or alkaline solution) on both the chloride and the sulfate interaction mechanisms. For instance, when C_3S pastes are immersed in the saturated lime solution (at pH 12.4), the amount of chlorides bound is nearly 3–4 times higher (whatever the equilibrium concentration considered) as compared to the amount bound with the alkaline solutions (at pH 13.5–13.6).

Similar results are found for the samples immersed in sulfate solutions. The comparison of chloride and sulfate interaction isotherms plotted in Fig. 3 is also interesting. Indeed, test results clearly show that the binding capacity of sulfate ions is much higher than that of chloride ions. For instance, for the immersion tests carried out using the lime-saturated solution, the amount of bound sulfate is nearly 3–5 times higher than the amount of bound chloride. The capacity of sulfates and chlorides to interact with the C_3S porous system seems then to be quite different.

Binding Isotherms: Hydrated Cement Pastes

The chloride and sulfate binding isotherms obtained on portland cement pastes are presented in Figs. 4 and 5, respectively. These figures show the amount of bound ions at various equilibrium concentrations (measured into the immersion solutions upon the completion of binding tests).

As for tricalcium silicate pastes, test results presented in Figs. 4 and 5 show that interaction mechanisms on cement paste systems are nonlinear in nature and can be modeled using a Freundlich isotherm (see Eq. 7). Binding experiment results also clearly show the influence of the chemical composition of the immersion solution (i.e., lime or alkaline solution) and the influence of the initial sample preparation (i.e., initially dried or saturated) on the chloride binding capacity of the cementitious systems. In all cases, the use of alkaline solutions as immersion solutions has reduced the amount of bound chloride by nearly 2–3 times. This reduction is, however, slightly less than the one observed for tricalcium silicate pastes (3–4 times). The initial water content of the material (i.e., dried at 11% RH or saturated at 100% RH) also has a significant effect on the ionic binding capacity of the material (see Fig. 4, graphs A and B). This effect is, however, clearly less pronounced at low chloride concentrations.

Test results obtained in this investigation emphasize the great binding

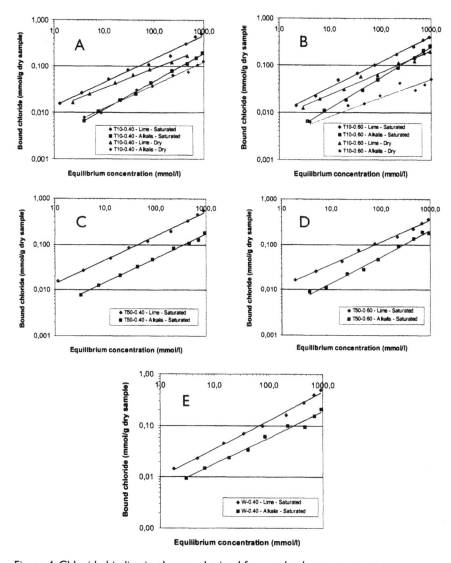

Figure 4. Chloride binding isotherms obtained for portland cement pastes.

capacity of hydrated cement pastes for sulfate ions. Indeed, the comparison of sulfate and chloride binding isotherms indicates that, for a given ionic concentration, the amount of bound sulfates is typically 4–8 times higher than the amount of bound chlorides. This effect has also been observed for both tricalcium silicate pastes.

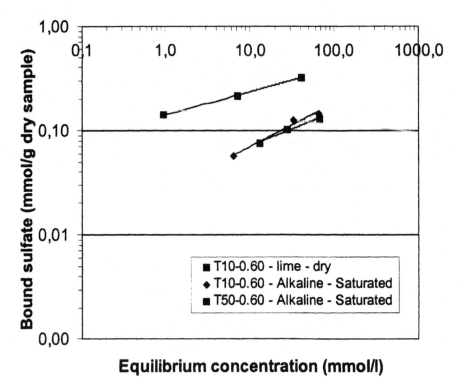

Figure 5. Sulfate binding isotherms obtained for hydrated portland cement pastes.

The results of Figs. 4 and 5 also indicate that the mineralogical composition of cement has a relatively weak influence (as compared to other parameters) on the binding capacity of the mixture. Data also suggest that the interaction of both chlorides and sulfates is determined much more by the total aluminate content of the cement than by its C_3A content. These results support the observations of Delagrave[3] and Zibara,[10] who say aluminum in C_4AF contributes to the binding capacity of the material.

The last parameter studied in this investigation was the effect of water/cement ratio on ionic binding (see Fig. 6). This figure gives the amount of bound chlorides on a unit mass of cement gel basis (calculated according to Eq. 6). The expression of bound chlorides as a function of the cement gel content, rather than on a unit mass of cement paste basis, is useful for comparing test results obtained on materials with different degrees of hydration (see Table V). Results shown in Fig. 6 indicate that the water/cement ratio does not have any significant effect on ionic binding.

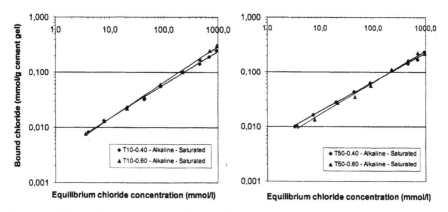

Figure 6. Effect of w/c ratio on chloride binding isotherms obtained for portland cement pastes.

Table V. Degree of hydration and maximum bound water content

Material	Degree of hydration	W_n^0
CSA T10 - w/c = 0.40	0.756	0.22
CSA T10 - w/c = 0.60	0.809	0.22
CSA T50 - w/c = 0.40	0.731	0.21
CSA T50 - w/c = 0.60	0.792	0.21

Numerical Simulations

In order to investigate the mechanisms of binding in hydrated cement systems, two series of numerical simulations were carried out. The first was performed to see if short-range electrical forces (resulting from the presence of electrical charges at the surface of cement-based materials) could be the origin of the interactions of chlorides and sulfates measured on tricalcium silicate pastes (see previous section). Those simulations were based on the theoretical model developed by Gouy and Chapman.[16,26]

The second series of numerical simulations was performed to establish if a minimal chloride or sulfate concentration is required to form new chloride or sulfate bearing phases (such as Friedel's salts or ettringite crystals) in neat cement paste systems used in this investigation. All simulations were carried out using the numerical model (named Simul) developed by Reardon.[21]

Simulation of Physical Binding Mechanisms in Hydrated Tricalcium Silicate Pastes

Description of the Numerical Model

As previously mentioned, a numerical model was based on a resolution of the Gouy and Chapman double-layer model. This model is based on the Poisson equation (Eq. 1) with an ionic concentration following a Boltzman distribution at the vicinity of pore wall.[16] The solution of the Poisson equation (with a Boltzman distribution) is given by:

$$\frac{\partial \phi}{\partial x} = \pm \left(\frac{2RT}{\varepsilon}\right)^{1/2} \left\{ \sum_i c_i (x = \infty) \left[\exp\left(\frac{-z_i F}{RT} \phi(x)\right) - 1 \right] \right\}^{1/2} \qquad (8)$$

This first-order differential equation was solved for four ionic species (Ca^{2+}, OH^-, Na^+, and Cl^-) using the well-known Runge-kutta numerical algorithm (fourth order). This algorithm was used to solve the Poisson equation over the whole double-layer thickness (i.e., from the Helmholtz plane [at $x = d$] up to the bulk solution [at $x = \infty$]). Numerical simulations were performed assuming that the Helmholtz electrical potential was close to the zeta potential measured experimentally.[27,28]

The amount of bound chloride or sulfate ions was determined assuming that all the ionic species included between the Helmholtz plane (at $x = d$) and the Debye-Huckel length (at $x = \kappa^{-1}$) were affected by the surface electrical potential and could then be considered as physically bound to the surface. The Debye-Huckel length was calculated according to Eq. 2. The zeta potential values used to perform the numerical simulations were based on the results of Viallis-Terrisse[8] and Nachbaur et al.[7] (see Fig. 2). The calcium concentrations required to evaluate the zeta potential were determined using a numerical equilibrium code (Simul).[21] Simulations were performed assuming that the investigated C_3S paste systems were in equilibrium with a saturated lime solution. The amount of bound chloride or sulfate ions was calculated according to the following equations:

$$\left[Cl^-\right]_{mol/g\ sample} = \rho\kappa^{-1} \int\limits_{x=d}^{x=\kappa^{-1}} \left(\left[Cl^-(x)\right]_{mmol/L} - \left[Cl^-(\kappa^{-1})\right]_{mmol/L} \right) dx \qquad (9)$$

and

Table VI. Input parameters used to perform the numerical simulations

Cement-based system	Salt	Concentration (mmol/)	Calcium content (mmol/)	Zeta potential (mV)	Specific surface (m2/g)*
C_3S + Ca(OH)$_2$	NaCl	0–500	21.1	32	200
C_3S + Ca(OH)$_2$	Na$_2$SO$_4$	0–50	21.1	32	200

*Typical value for dried C_3S pastes, see Ref. 29.

$$\left[SO_4^{2-}\right]_{mol/g\ sample} = \rho\kappa^{-1} \int\limits_{x=d}^{x=\kappa^{-1}} \left(\left[SO_4^{2-}(x)\right]_{mmol/L} - \left[SO_4^{2-}\left(\kappa^{-1}\right)\right]_{mmol/L}\right) dx \quad (10)$$

where $[Cl^-(x)]$ and $[SO_4^{2-}(x)]$ are the chloride and sulfate concentration distributions extending from the surface (at $x = d$) up the Debye-Huckel length (at $x = \kappa^{-1}$), and ρ is the specific surface (m^2/g) of the investigated system.

Initial conditions for each simulation performed using the Gouy and Chapman double layer model are given in Table VI.

Numerical Results

Typical numerical results are presented in Fig. 7. This figure shows the evolution of the chloride concentration in the vicinity of the charged surface for various sodium chloride concentrations in the bulk solution. Numerical results clearly show a marked increase in the chloride ion concentration near the solid/liquid interface. This increase of the chloride concentration contributes to restore the electroneutrality of the system.

The amount of bound chloride and sulfate ions calculated using Eqs. 9 and 10 is shown in Fig. 8. All chloride and sulfate numerical binding isotherms presented in this figure have a nonlinear shape similar to the one obtained for the hydrated tricalcium silicate pastes (see the previous section). Analysis indicates that the numerical results (using the Gouy and Chapman double-layer model with the input parameters presented in Table VI) are in reasonable agreement (considering the uncertainty related to the specific surface evaluation) with experimental data obtained on C$_3$S pastes.

However, for sulfate ions, the numerical and experimental binding isotherms do not correlate as well. It is believed that this phenomenon can be attributed to the specific adsorption of sulfate ions on the C-S-H surface. The discrepancy between the numerical and experimental results shown in

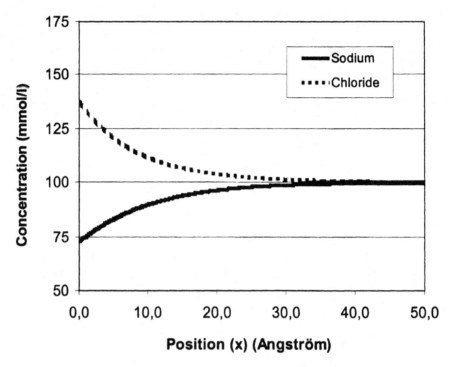

Figure 7. Evolution of chloride concentration at the vicinity of the charged surface at various sodium chloride concentrations in the bulk solution [C_3S + $Ca(OH)_2$ + NaCl system].

Fig. 8(b) suggests that surface complexation phenomena (which are not accounted for by the Gouy and Chapman double-layer model) contribute significantly to the binding of sulfate ions in hydrated C_3S paste. Electrostatic interaction also appears to be marginal as compared to the specific adsorption of sulfate ions. This situation will be further discussed below.

Determination of Chloride and Sulfate Threshold Concentrations to Form New Compounds

The second series of numerical simulations carried out in this investigation was performed to establish if a minimal chloride or sulfate concentration is required to form new chloride or sulfate bearing phases (such as Friedel's salts or ettringite crystals) in neat cement paste systems. All simulations were performed using the numerical code (named Simul) developed at the University of Waterloo (Canada) by Reardon.[21]

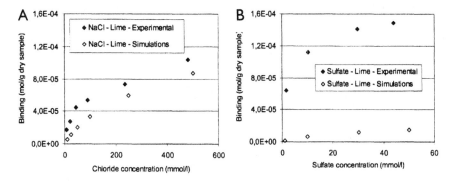

Figure 8. Predicted and experimental binding isotherms obtained for the C_3S + $Ca(OH)_2$ + NaCl, C_3S + NaOH + KOH + NaCl, and C_3S + $Ca(OH)_2$ + Na_2SO_4 systems.

Chemical Equilibrium Model

Numerical simulations were performed assuming that 0.40 w/c neat cement pastes (prepared with one of the three portland cements shown in Table II) were immersed in 50-mL lime or alkaline solutions. Sodium chloride or sodium sulfate was then progressively added to each investigated system. The hydrated phases assumed to be in equilibrium with the immersion solution were calcium silicate hydrates, portlandite, ettringite, brucite, hydrogarnet, and (if thermodynamically stable) Friedel's salt. Solubility constants of these hydrated phases are given in Table I. All these values (except the one for Friedel's salts) are those proposed by Reardon.[21] The solubility of Friedel's salts presented in Table I is an average value obtained from test results published by Abate and Scheetz[30] and Damidot et al.[22] Chemical activity was also considered in the numerical simulations. Activity coefficients were calculated using Pitzer's equations.

Numerical Results

For each amount of sodium chloride added to the lime or alkaline immersion solutions, the chemical stability of all hydrated phases and the possible formation of Frieldel's salt were systematically verified. The minimum chloride concentration required to stabilize the Friedel's salt was called the chloride threshold level. Numerical results for Friedel's salt formation are given in Table VII. For all cement paste systems, numerical results clearly indicate the existence of a minimal chloride concentration to form Friedel's salts. Numerical results also reveal that the chloride threshold level is markedly dependent on the immersion solution composition.

Table VII. Equilibrium chloride concentrations required to stabilize Friedel's salts in the cement paste systems investigated

Material	Immersion solution	Equilibrium chloride concentration (mmol/L)
CSA T10 - w/c = 0.40	Ca(OH)$_2$ (saturated solution)	27.5
	NaOH (195 mmol/L) + KOH (595 mmol/L)	375
CSA T50 - w/c = 0.40	Ca(OH)$_2$ (saturated solution)	24.5
	NaOH (170 mmol/L) + KOH (415 mmol/L)	278
White - w/c = 0.40	Ca(OH)$_2$ (saturated solution)	17.2
	NaOH (130 mmol/L) + KOH (30 mmol/L)	71.0

In the case of the sodium sulfate addition, no Frieldel's salt precipitation was obviously considered. For all cement paste systems studied, numerical results show that all hydrated phases (i.e., calcium silicate hydrates, portlandite, brucite, hydrogarnet) as well as ettringite were stable even if no sodium sulfate was initially added to the lime or alkaline immersion solutions. The stability of ettringite comes from the initial sulfur content of portland cement. Numerical results clearly show that ettringite can precipitate even if the sodium sulfate salt concentration added to the immersion solutions is minimal.

Discussion

Test results obtained in this investigation have clearly emphasized the significant influence of the composition of the immersion solutions (i.e., lime or alkaline solutions) on ionic binding mechanisms for all the systems studied. In the case of hydrated tricalcium silicate pastes, the influence of the immersion solutions on chloride binding appears to be controlled predominantly by short-range electrical forces. Indeed, the high calcium concentration in the C$_3$S-Ca(OH)$_2$-H$_2$O system (as compared to the C$_3$S-NaOH-KOH-H$_2$O system) results in an important surface potential (or zeta potential) at the system liquid/solid interface. The influence of the calcium concentration on zeta potential has clearly been shown by Viallis-Terrisse[8] (see Fig. 2). With an important surface potential (nearly +30 mV for saturated lime solutions) chloride ions tend to be tightly bound to the surface. These chloride ions are then considered to be physically bound. The good agreement between the numerical and experimental results presented in Fig. 8 tends to confirm the importance of short-range electrical forces on chloride binding in C_3S paste systems immersed in saturated lime solutions. This good correlation between the two series of data also supports the hypothesis of Viallis-Terrisse[8] and Nachbaur et al.[7] according to which chloride ions are not specifically adsorbed on the surface of hydrated cement systems.

In the presence of alkaline solutions, the calcium concentration of the hydrated C_3S system pore solution is, however, markedly reduced. Indeed, results of numerical simulations carried out with the chemical equilibrium code have shown that the calcium concentration is typically reduced from 21.1 to 1.19 mmol/L when the alkaline content of the solution is increased from 0 to 450 mmol/L. At low calcium contents, the zeta potential is significantly reduced and physical binding should then be low. The comparison of chloride binding isotherms obtained for C_3S-$Ca(OH)_2$-H_2O and C_3S-NaOH-KOH-H_2O systems (see Fig. 3) seems to support this hypothesis.

Test results also emphasize the importance of the solution composition on the amount of sulfates bound by the hydrated C_3S pastes. The poor correlation between the numerical and experimental binding isotherms presented in Fig. 8 suggests that ionic interactions are not limited solely to short-range electrical forces. In fact, as pointed out by Nachbaur et al.,[7] sulfate ions are most probably specifically adsorbed on the C-S-H surface. This hypothesis is supported by zeta potential measurements, which clearly show a reduction of surface electrical potential when sulfate ion concentration is increased.[28] It is believed that sulfate ionic binding mechanisms in tricalcium silicate pastes are attributed mainly to a combination of physical binding (short-range electrical forces) and surface chemical reaction (sulfate specifically adsorbed).

Numerical results presented in Table VII indicate that a minimal chloride concentration is required to form new chloride compounds (i.e., Friedel's salts) in neat cement paste systems. The chloride threshold values were found to be influenced by the ionic strength of the immersion solutions. More precisely, the alkali content seems to play a major role in the chemical stability of Frieldel's salts. As pointed out by Damidot et al.,[22] the minimum chloride concentration required to stabilize calcium monochloroaluminate is markedly increased when sodium or potassium ions are added to the solution. For instance, Damidot et al. have calculated that the lowest chloride concentration required to stabilize Friedel's salts increases by nearly ten times when the Na_2O content of the CaO-Al_2O_3-$CaCl_2$-Na_2O-H_2O

systems goes from 0 to 250 mmol/L. The marked influence of the alkali content on Friedel's salt precipitation appears to be an important mechanism at the origin of the higher chloride binding capacity of cement pastes immersed in saturated lime solutions.

As suggested by zeta potential measurements, the tendency of hydrated cement systems to bind more chlorides when immersed in lime-saturated solutions is probably related to the ability of these ions to physically interact with the matrix. Indeed, the test results presented in Fig. 4 clearly show that chloride ions can be bound to portland cement pastes even at chloride concentrations lower than the threshold level presented in Table VII. This appears to be a clear indication that chloride binding in portland cement pastes is not limited solely to the formation of new chloride-bearing phases. It is believed that, at low chloride concentrations, chloride binding is mainly dominated by short-range electrical forces.

In that respect, a comparison of chloride binding isotherms obtained on hydrated tricalcium silicate paste and portland cement paste (see Fig. 9) clearly shows that the amount of bound chloride is almost the same for both systems at low chloride concentration. When the chloride concentration reaches 30–40 mmol/L, the amount of bound chloride is then slightly higher for portland cement paste as compared to C_3S paste. According to our previous analysis, this effect is believed to be caused by Friedel's salt formation. However, even at high chloride concentrations, chloride binding is still quite important in the C_3S paste system as compared to the cement paste system studied.

Numerical simulations have also been performed to establish if the formation of new sulfate-bearing phases (such as ettringite) required a minimal sulfate concentration. Results indicate that there is no minimal sulfate concentration required to chemically bind sulfate ions, whatever the immersion solution used. Numerical simulations reveal that sulfate ions were consumed to produce ettringite. No precipitation of gypsum was predicted. The main conclusion of this analysis is that sulfate ions are used to form ettringite even if the amount of sulfate ions added to the immersion solution is as low as 1 mmol/L. This conclusion appears to be independent of the alkaline content of the immersion solutions. It is also believed that chemical binding (or ettringite formation) is the dominant interaction mechanism when sulfate ions are in contact with portland cement pastes.

The last hypothesis is supported by the numerical results presented above. Indeed, results have shown that the amount of sulfate ions physical-

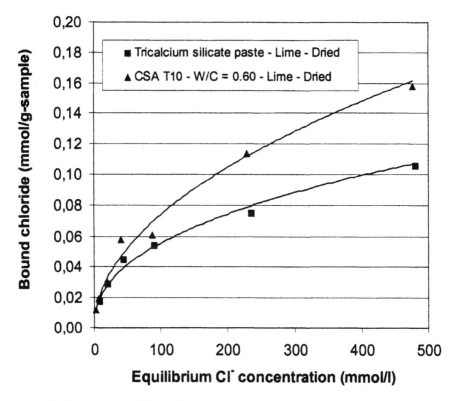

Figure 9. Comparison of chloride binding isotherms in tricalcium silicate past (w/c = 0.50, initially dried) and portland cement paste system (CSA T10, w/c = 0.60, initially dried.

ly bound to the hydrated tricalcium silicate paste was relatively low. For instance, for $[SO_4^{2-}] = 30$ mmol/L, calculations indicate that amount of sulfate ions physically bound to the system is limited to 0.012 mmol/g (see Fig. 8) for a total amount of sulfates of 0.295 mmol/g bound to a 0.60 w/c portland cement paste immersed in the saturated lime solution (see Fig. 10). This difference suggests that physical binding (in presence of sulfate ions) only plays a marginal role. It should also be pointed out that the specific adsorption of sulfate ions on the C-S-H surface of hydrated cement paste systems is also considered to be minimal. Indeed, since hydrated cement pastes naturally contain sulfates at concentrations typically ranging from 1 to 15 mmol/L, it is reasonable to assume that many of these ions are already complexed at the surface of the solid. Consequently, almost no

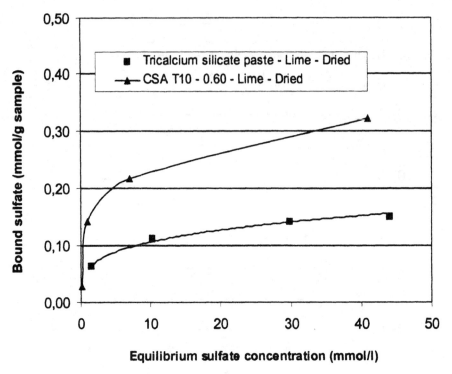

Figure 10. Comparison of sulfate binding isotherms in hydrated tricalcium silicate paste (w/c = 0.50, initially dried) and the hydrated portland cement system (CSA T10, w/c = 0.60, initially dried).

external sulfate ions can be specifically adsorbed on the C-S-H surface in portland cement pastes.

A typical comparison of sulfate binding isotherms obtained on tricalcium silicate paste and portland cement paste is given in Fig. 10. The comparison of these two isotherms clearly shows a marked difference (even at very low sulfate concentrations) in the amount of bound sulfates. This difference is believed to be the consequence of the important chemical binding (i.e., ettringite formation) in the hydrated portland cement system.

Test results obtained in this investigation show that water/cement ratio and the portland cement type do not have a marked influence on ionic binding. The slight influence of cement type is quite surprising since the aluminum contents of the three cements used are relatively different (see Table II). It is reasonable to assume that the cement with the higher aluminate content (CSA Type 10) should be the one with the higher binding capacity, and it is interesting to point out that this cement is also the one with the highest alkaline content. On the opposite side, the cement with the lower aluminate content (white cement) is also the binder with the lower alkaline content. According to the experimental and numerical results presented previously, it can be argued that the expected high binding capacity of CSA Type 10 cement (with its high aluminum content) is probably counterbalanced by the high alkaline content of this cement. It has previously been clearly shown that alkaline ions tend to markedly reduce ionic binding. Both aluminate and alkaline content of portland cement should then be considered in regard to ionic binding capacity of cement-based materials.

Finally, sample preparation has been found to play an important role in chloride and sulfate binding. The two test conditions used in this investigation (initially dried using silica gel and initially saturated) have markedly influenced ionic binding. The lower binding capacity of initially dried portland cement paste systems can probably be attributed to a reduction of the material specific area. The effect of drying on the material specific surface has been investigated by Rarick et al.[29] Test results obtained by these authors indicate that the specific surface of the dried materials can be sensibly reduced. With an apparent lower specific surface, it is believed that the potential surface reaction sites of dried cement pastes are reduced as well as possible amount of bound ions.

Conclusion

Experimental and numerical results obtained using chloride ions have shown that physical binding (i.e., short-range electrical forces) appears to be the dominating interaction mechanism in tricalcium silicate pastes. In portland cement pastes exposed to chloride solutions, physical binding seems to be the main interaction phenomenon at low chloride concentrations (i.e., lower than the threshold chloride content required to stabilize Frieldel's salt). At high chloride concentrations, Frieldel's salt can be formed, and then both chemical and physical binding should be considered as significant interaction mechanisms. It is, however, important to underline that the required chloride concentration to get chemical binding is closely related to the chemical composition of cement paste immersion solutions.

For sulfate ions, experimental results have shown that chemical binding (i.e., ettringite formation) appears to be the main binding mechanism in portland cement pastes. Numerical analysis has indicated that no minimal sulfate concentration was required to form ettringite in the systems investigated. Test results obtained on tricalcium silicate pastes seem to indicate that sulfate ions could be specifically adsorbed on C-S-H surface. The same results have finally shown that physical binding plays only a marginal role in the total binding mechanism.

Results obtained in this investigation have finally indicated that (1) the amount of bound sulfate is markedly higher than the amount of bound chloride, (2) cement chemical compositions tested and water/cement ratio have only a slight influence on ionic binding, and (3) sample preparation (i.e., initially dried or saturated) has a significant effect on the total ionic interactions of portland cement pastes tested.

References

1. V. S. Ramachandran, "Possible States of Chloride in the Hydration of Tricalcium Silicate in the Presence of Calcium Chloride," *Mater. Struct.,* **4** [19] 3–12 (1971).
2. M. Castellote, C. Andrade, and C. Alonso, "Chloride-Binding Isotherms in Concrete Submitted to Non-Steady-State Migration Experiments," *Cem. Concr. Res.,* **29,** 1799–1806 (1999).
3. A. Delagrave, "Mécanismes de pénétration des ions chlore et de dégradation des systèmes cimentaires normaux et à haute performance," Ph.D. thesis, Laval University, Québec, 1996.
4. J. J. Beaudoin, V. S. Ramachandran, and R. F. Feldman, "Interaction of Chloride and C-S-H," *Cement Concr. Res.,* **20** [6] 875–883 (1990).
5. L. Divet and R. Randriambololona, "Delayed Ettringite Formation: The Effect of Temperature and Basicity on the Interaction of Sulfate and C-S-H Phase," *Cem. Concr. Res.,* **28,** 357–363 (1998).
6. O. Wowra and M. J. Setzer, "Sorption of Chlorides on Hydrated Cements and C_3S Pastes"; pp. 146–153 in *Frost Resistance of Concrete.* Edited by Setzer and Auberg. E&FN Spon, London, 1997.
7. L. Nachbaur, P. C. Nkinamubanzi, A. Nonat, and J. C. Muttin, "Electrokinetic Properties Which Control the Coagulation of Silicate Cement Suspensions during Early Age Hydration," *J. Colloid Interface Sci.,* **202** [2] 261–268 (1998).
8. H. Viallis-Terrisse, "Interaction des silicates de calcium hydratés, principaux constituants du ciment, avec les chlorures d'alcalins: Analogie avec les argiles," Ph.D. thesis, UFR des Sciences et Techniques, Université de Bourgogne, 2000.
9. J. Arsenault, "Étude des mécanismes de transport des ions chlore dans le béton en vue de la mise au point d'un essai de migration," Ph.D. thesis, Laval University, Québec, 1999.
10. H. Zibara, "Binding of External Chlorides by Cement Pates," Ph.D. thesis, University of Toronto, 2002.
11. E. Samson, "Modélisation numérique du transport ionique par diffusion dans les matériaux poreux," Ph.D. thesis, Laval University, Québec, 2004.

12. S. Catinaud, Ph.D. Thesis, Laval University, Québec, 2001.
13. S. Y. Hong and F. P. Glasser, "Alkali Binding in Cement Pastes Part I. The C-S-H Phase," *Cem. Concr. Res.,* **29**, 1893–1903 (1999).
14. E. Nägele, "The Zeta-Potential of Cement," *Cement Concr. Res.,* **17**, 573–580 (1987).
15. T. Zhang and O. E. Gjorv, "Diffusion Behavior of Chloride Ions in Concrete," *Cem. Concr. Res.,* **26** [6] 907–917 (1996).
16. J. O'M. Bockris, B. E. Conway, and E. Yeager, *Comprehensive Treatise of Electro-Chemistry, Volume 1: The Double Layer.* Plenum Press, 1980.
17. J. Rubin, "Transport of Reacting Solutes in Porous Media: Relation between Mathematical Nature of Problem Formulation and Chemical Nature of Reactions," *Water Resources Res.,* **19** [5] 1231–1252 (1983).
18. H. F. W. Taylor, "Ettringite in Cement Paste"; in *Proceedings of the RILEM Conference Concrete — From Material to Structure* (Arles, France, September 11–12, 1996).
19. S. Kumar and R. Kameswara, "Strength Loss in Concrete Due to Varying Sulfate Exposures, Cement and Concrete Research," **25** [1] 57–62 (1995).
20. J. R. Clifton and J. M. Pommersheim, "Sulfate Attack of Cementitious Materials: Volumetric Relations and Expansions." NISTIR 5390, 20 pp. National Institute of Standards and Technology, 1994.
21. E. J. Reardon, "Problems and Approaches to the Prediction of the Chemical Composition in Cement/Water Systems," *Waste Management,* **12** (1992).
22. D. Damidot, U. A. Birmin-Yauri, and F. P. Glasser, "Thermodynamic Investigation of the CaO-Al$_2$O$_3$-CaCl$_2$-H$_2$O System at 25°C and the Influence of Na$_2$O," *Il Cemento,* **4**, 243–254 (1994).
23. L. Tang and L. O. Nilsson, "Chloride Binding Capacity and Binding Isotherms of OPC Pastes and Mortars," *Cem. Concr. Res.,* **23** [2] 247–253 (1993).
24. J. Marchand, B. Gérard, and A. Delagrave, "Ion Transport Mechanisms in Cement-Based Materials"; pp. 307–400 in *Materials Science of Concrete V.* American Ceramic Society, Westerville, Ohio, 1995.
25. O. Mejlhede Jensen, M. S. H. Korzen, H. J. Jakobsen, and J. Skibsted, "Influence of Cement Constitution and Temperature on Chloride Binding in Cement Paste," *Adv. Cem. Res.,* **12** [2] 261–268 (2000).
26. A. W. Adamson, *A Textbook of Physical Chemistry,* 3rd ed. Academic Press College, 1986.
27. R. J. Hunter, *Foundations of Colloid Science,* 2nd ed. Oxford University Press, 2001.
28. P. Henocq, "Modélisation des interaction ioniques à la surface des Silicates de Calcium Hydratés," Ph.D. thesis, Laval University, Québec, 2004.
29. R. L. Rarick, J. I. Bhatty, and H. M. Jennings, "Surface Area Measurement Using Gas Sorption: Application to Cement Paste"; pp. 1–39 in *Materials Science of Concrete IV.* Edited by J. P. Skalny. American Ceramic Society, Westerville, Ohio, 1994.
30. C. Abate and B. E. Scheetz, "Aqueous Phase Equilibra in the System CaO-Al$_2$O$_3$-CaCl$_2$-H$_2$O: The Significance and Stability of Friedel's Salt," *J. Am. Ceram. Soc.,* **78** [4] 939–944 (1992).

Mechanisms of Frost Damage

George W. Scherer and John J. Valenza II

The mechanisms responsible for damage from internal freezing and salt scaling are reviewed. The primary cause of internal damage is crystallization pressure, and the role of the air voids is to provide sites for nucleation of macroscopic ice. The thermodynamics of the stress development are reviewed, and the predicted pressures are shown to be in quantitative accord with measured contraction of freezing bodies (including porous glass and cement paste). The origin of salt scaling damage is less clear. We examine two mechanisms that seem to account for most of the experimental observations: the bimaterial effect (i.e., thermal expansion mismatch between ice and cement paste) and salt-induced swelling. A sensitive experiment, in which a layer of water is frozen on top of a thin plate of cement paste and the deflection of the plate is measured, reveals the large stresses produced by these mechanisms. Cracking of the ice layer is promoted by brine pockets, and this may account for the pessimum concentration for scaling damage. Salt-induced swelling, which seems to result from a combination of crystallization pressure and ion exchange, also contributes to the superficial stresses and may exacerbate scaling.

Introduction

This chapter reviews the evidence for the mechanisms responsible for damage to concrete during freeze/thaw cycles. We will examine the relative importance of hydraulic pressure and crystallization pressure in creating internal stresses and cracking. The evidence favors crystallization pressure as the primary cause of damage, so we will take a detailed look at the thermodynamic rules governing the propagation of ice through the porous network, and identify the factors controlling the magnitude of the stresses. The mechanisms responsible for salt scaling are quite different. We will review the phenomenology and the ideas that have been proposed to account for scaling, then will describe some recent experiments that shed new light on the processes responsible for salt scaling.

Internal Crystallization

Powers[1] argued that harmful stresses could result from hydraulic pressure created by the volume increase as water transforms to ice. He proposed that suitably spaced air voids would prevent the development of excessive pressure, and predicted the necessary spacing for entrained air voids. It is a common misconception that freeze/thaw damage is entirely due to hydraulic pressure, and that adequate air entrainment is all that is required for complete protection. In fact, dilation of a sample of concrete occurs during freezing when the pores contain an organic liquid, which does not expand as it crystallizes[2,3]; this is clear evidence of the importance of crystallization pressure, since expansive hydraulic pressure is not created in this case. An elegant study by Powers and Helmuth[4] revealed that properly air-entrained concrete still suffers stress and dilatation caused by crystallization pressure. This is stress created by repulsion (or, disjoining pressure) between the ice crystal and the minerals in the pore walls. It is a peculiarity of ice that the solid phase has a lower refractive index, as well as lower density, than the liquid. Therefore, at the ice/water/mineral interface, the refractive index of the liquid lies between those of the solid phases; in such a case, the van der Waals forces between the solids are repulsive.[5] Additional contributions to the disjoining pressure, P_D, may come from electrostatic forces or the structuring of the solvent at the solid surface. It has been demonstrated[6] that the magnitude of this repulsion is so large that the pressure required to push ice into contact with the pore wall exceeds the strength of concrete. Therefore, a liquid film always separates the ice and the pore wall. The presence of the film permits the crystal to grow while pushing away objects. Corte[7] showed that a unidirectionally frozen column of ice could lift a heavy overburden of soil particles. As he noted, this requires molecules of water to enter the gap between the ice and the soil particles, and attach to the growth surface of the ice. This is the mechanism of frost heave,[8] and the same repulsive force is exerted on the pore walls as ice grows in concrete.

Powers and Helmuth[4] showed that concrete without air entrainment expands as it is cooled, and attributed this to hydraulic pressure produced during freezing. As indicated in Fig. 1 (dashed curve), the dilatation stops almost immediately during an isothermal hold; the dilatation begins again when cooling is resumed. In contrast, the sample with adequate air entrainment (solid curve) contracts as it is cooled. The shrinkage is attributed to the growth of ice crystals inside the air voids, which suck water from the capillary pores and create negative pressure in the liquid; when cooling is arrested, the contraction diminishes slowly. Helmuth and Powers interpreted the latter effect as resulting from the melting of ice crystals that had nucleated in the mesopores: the strain relaxes as the liberated water flows to the air voids and freezes there. The driving force for the transfer of water from the ice in mesopores to the ice in the air voids is the elevated chemical potential created by the highly curved surfaces on the smaller crystals.

Powers and Helmuth argued that damage could result from local strain around those crystals, even though the body exhibits overall contraction. Helmuth[9] later concluded that frost damage was entirely attributable to crystallization pressure, as discussed below. In contrast, Fagerlund[10] argues persuasively that freezing damage results primarily from hydraulic pressure, exacerbated by reductions in permeability as pores fill with ice. We will examine these phenomena in turn, and evaluate their relative importance to frost damage.

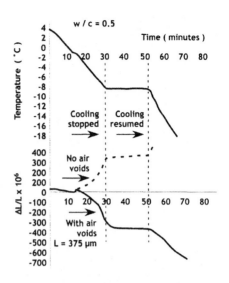

Figure 1. Dilatation during freezing of concrete with air entrainment (air void spacing $L = 375$ μm, solid curve) and without it (dashed); both samples have 45% porosity. (Redrawn from Ref. 4.)

Crystallization Pressure

The magnitude of the stress exerted by crystals in the capillary pores can be quantified as follows. The thermodynamics of this problem have been studied for a century[11] and the principles have been discussed by several authors.[6,12-14] (Some of the treatments of this problem in the literature provide the right results, even when the method of analysis lacks rigor. The correct approach is reviewed in the appendix.) Let S_L and S_C be the molar entropies of the liquid and crystal, respectively; the molar entropy of fusion is $\Delta S_{fv} = (S_L - S_C)/v_C$, where v_C is the molar volume of the crystal. If T_m is the melting point, then the chemical potential driving growth of the crystal at temperature T is given approximately by $\Delta \mu \approx \Delta T \, \Delta S_{fv}$. If the crystal encounters an obstruction (such as a pore wall), the crystal sustains a film of liquid between itself and the obstruction and continues trying to grow, pushing the obstacle away. The pressure that ice can generate in this way is roughly 1.2 MPa per degree of undercooling,[14] so it requires only a few degrees of undercooling to produce stresses that exceed the tensile strength of concrete.

Consider the crystal in Fig. 2, which is at equilibrium in a cylindrical pore with radius r_p; the temperature has been adjusted so that the crystal

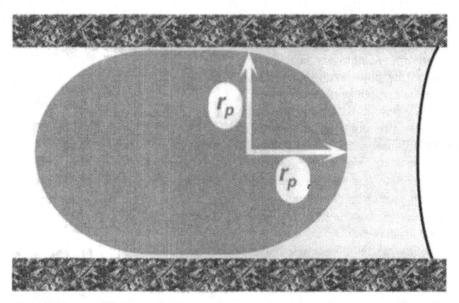

Figure 2. Crystal of ice in cylindrical pore with radius r_p. The hemispherical end of the crystal has curvature $\kappa_{CL} = 2/(r_p - \delta)$, where δ is the thickness of the unfrozen film, but the cylindrical side has curvature $\kappa_{CL} = 1/(r_p - \delta)$.

neither grows nor shrinks. According to the Gibbs-Thomson equation,[15] this equilibrium temperature is (see Eq. 47 of the appendix):

$$T = T_m - \frac{\gamma_{CL} \kappa_{CL}}{\Delta S_{fv}} \tag{1}$$

where T_m is the melting point of a large flat crystal, γ_{CL} is the specific energy of the crystal/liquid interface, and κ_{CL} is the curvature of the interface. For a spherical crystal, the melting point is

$$T = T_m - \frac{2\gamma_{CL}}{(r_p - \delta)\Delta S_{fv}} \tag{2}$$

where $\kappa_{CL} = 2/(r_p - \delta)$ and δ is the thickness of the liquid film between the crystal and the pore wall; for water, $\delta \approx 0.9$ nm.[16] For ice, $\gamma_{CL} \approx 0.04$ J/m^2 and $\Delta S_{fv} \approx 1.2$ J/(cm$^3 \cdot$K),[16] so the melting point of ice is reduced by 2°C when $r_p \approx 33$ nm, by 5°C when $r_p \approx 13$ nm, and by 10°C when $r_p \approx 7$ nm.

Although the hemispherical ends of the crystal in Fig. 2 have curvature $\kappa_{CL}^E = 2/(r_p - \delta)$, the cylindrical sides have curvature $\kappa_{CL}^S = 1/(r_p - \delta)$. Since both surfaces are at the same temperature, additional pressure, P_A, must be applied to the cylindrical sides to prevent growth; this pressure is supplied by the pore wall. For a crystal of ice in pure water, the equilibrium condition is given by Eq. 45 of the appendix:

$$v_C\left(P_A + \gamma_{CL}\kappa_{CL}\right) = \left(v_L - v_C\right)\left(P_L - P_e\right) + \left(S_L - S_C\right)\left(T_m - T\right) \tag{3}$$

Laplace's equation relates the pressure in the liquid to the curvature of the liquid/vapor interface, κ_{LV}[15]:

$$P_L - P_e = \gamma_{LV}\kappa_{LV} \tag{4}$$

where γ_{LV} is the surface tension of the liquid/vapor interface. Thus, Eq. 3 can be written as

$$v_C\left(P_A + \gamma_{CL}\kappa_{CL}\right) = \left(v_L - v_C\right)\gamma_{LV}\kappa_{LV} + \left(S_L - S_C\right)\left(T_m - T\right) \tag{5}$$

For the crystal in Fig. 2, the relative humidity* (RH) and temperature are such that the right side of Eq. 5 is in balance with the hemispherical end of the crystal, where $P_A = 0$ and the curvature is κ_{CL}^E. Equilibrium is preserved on the sides of the crystal if $P_A + \gamma_{CL} \kappa_{CL}^S = \gamma_{CL} \kappa_{CL}^E$; that is, the pore wall must impose a radial pressure on the crystal given by

$$P_A = \gamma_{CL}\left(\kappa_{CL}^E - \kappa_{CL}^S\right) \tag{6}$$

The highest pressure is generated in a large pore with small entries, as in Fig. 3. At point P, where the pore wall is relatively flat, the crystal has a large positive radius of curvature, so κ_{CL}^S is small. Where the pore wall protrudes at point N, the crystal adopts a negative radius of curvature, so the pressure on the wall is very high; the upper bound on P_A is the disjoining pressure, P_D. In the cylindrical pore of Fig. 2, $P_A = \gamma_{CL}/(r_p - \delta)$.

---

*The relative humidity in air voids within the body is fixed by the ice, so it cannot be independently controlled. The same is true at the exterior surface of the body in a closed system (such as a cold chamber in a laboratory). In an open system, if the RH is higher than the value in equilibrium with ice at the existing temperature, then moisture will condense on the ice and in capillaries; the growth rate of the macroscopic ice will be equal to the rate of condensation. If the RH is below the equilibrium value, ice and water will evaporate.

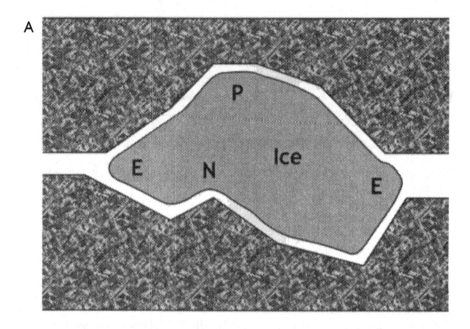

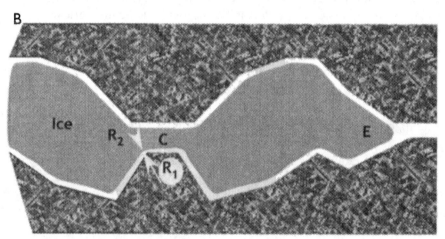

Figure 3. (a) A crystal of ice in a large pore with small entries has a small positive radius of curvature at the pore entries (E). The curvature elsewhere on the surface of the crystal is dictated by the shape of the pore wall. At P, there is a large positive radius of curvature, whereas at N there is a small negative radius of curvature. The pressure, P_A, exerted on the pore wall varies from point to point such that Eq. 6 is obeyed. (b) The constriction at C has a negative radius of curvature, R_1, in the plane of the page, but a positive radius of curvature, R_2, in the plane perpendicular to the page. Therefore, the net curvature at constrictions, $\kappa_{CL} = 1/R_1 + 1/R_2$, may be small; in such cases, $P_A \approx \gamma_{CL}\kappa_{CL}^E$ at constrictions.

For either a cylindrical[6] or spherical pore,[14] the radial pressure creates a tensile hoop stress in the wall roughly equal to:

$$\sigma_\theta \approx \frac{\gamma_{CL}}{2}\left(\kappa_{CL}^E - \kappa_{CL}^S\right)$$

(7)

For example, assuming 100% RH (so $\kappa_{LV} = 0$), ice could enter a pore with radius $r_p = 15$ nm at $-5°C$, at which point $\Delta S_{fv}\Delta T = 6$ MPa. Eq. 7 indicates that the crystal would exert a stress of $\sigma_\theta \approx 3$ MPa on the pore wall. Thus, the crystallization stress approaches the tensile strength of concrete in small pores, which freeze at low temperatures.

Condition for Penetration of a Pore

According to Eq. 1, the ice penetrates any pore whose entrance is large enough to satisfy the following condition:

$$r_p \geq \delta + \frac{2\gamma_{CL}}{(T_m - T)\Delta S_{fv}}$$

(8)

The crystallization stress grows as the temperature drops, permitting ice to penetrate smaller and smaller pores. Let us examine more carefully the way this invasion progresses.

If an ice crystal nucleates at the exterior surface or in an air void, it can grow without generating any crystallization pressure (unless it becomes so large that it fills the void). It cannot penetrate the surrounding capillary pores until the temperature is low enough to satisfy Eq. 1, but until that happens the crystal will suck water from the surrounding capillaries and create negative pressure in the pores. As indicated in Fig. 4, the crystal in the void (shown at the top of the figure) is in contact with the atmosphere in the void at atmospheric pressure P_e, and with the liquid in the pore at pressure P_L. The crystal bulges into the pore, but it cannot enter, even though the temperature is below the melting point, because T is too high to satisfy Eq. 1. The positive curvature, κ_{CL}^E, at the "nose" of the bulging crystal/liquid interface (where $P_A = 0$) is found from Eq. 5:

$$\gamma_{CL}\kappa_{CL}^E = \left(\frac{v_L - v_C}{v_C}\right)\gamma_{LV}\kappa_{LV} + \left(\frac{S_L - S_C}{v_C}\right)(T_m - T)$$

(9)

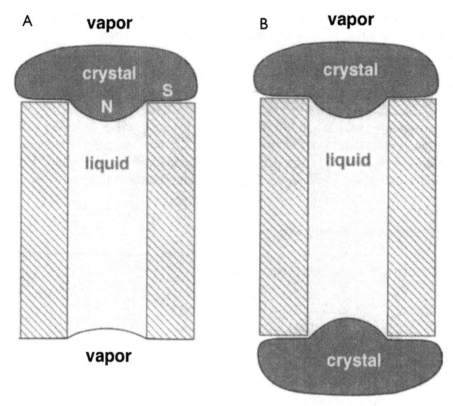

Figure 4. A crystal in an air void is in contact with the vapor in the void at pressure P_e and with the pore liquid at pressure P_L. The negative pressure in the liquid can be sustained by an opposing liquid/vapor interface (a) or crystal/liquid interface (b). The curvature is different at the nose (N) and the shoulder (S) of the crystal, but the difference is offset by the pressure exerted by the wall, P_A.

The other side of the macroscopic crystal is also in equilibrium with the liquid (that is, the vapor pressure of the water is equal to that of macroscopic ice). Therefore, the pressure in the liquid is (see Eq. 52 of the appendix):

$$v_L(P_L - P_e) = -(S_L - S_C)(T_m - T) = v_L \gamma_{LV} \kappa_{LV} \qquad (10)$$

where the second equality follows from Laplace's equation (Eq. 4). Using Eq. 10 to eliminate the entropic term from Eq. 9, we obtain a relationship between the curvatures[12,14]:

$$\gamma_{CL}\kappa_{CL}^E = -\gamma_{LV}\kappa_{LV} \qquad (11)$$

Since the curvature at the nose of the crystal is positive, the pressure in the liquid must be negative. The pressure can be sustained by another crystal, as in Fig. 4(b), as well as by a liquid/vapor interface at the other end of the pore, as in Fig. 4(a). For water, $\gamma_{CL} < \gamma_{LV}$, so the curvature of the liquid/vapor interface is always less than that of the crystal liquid interface, which means that the ice cannot drain the liquid from a cylindrical pore; if the temperature becomes low enough, the liquid will suck the ice into the pore. However, if the pore is tapered, with the small end near the ice, the crystal will be able to drain most of the liquid from the pore. This means that it would be favorable to have small pores lining air voids, because this would enable ice to drain water from more of the surrounding pores in the paste.

For either of the cases shown in Fig. 4, we find from Eq. 10 that the pore pressure is related to the temperature by

$$P_L - P_e = -\left(\frac{S_L - S_C}{v_L}\right)(T_m - T) \tag{12}$$

(In this case, because the ice is at atmospheric pressure, the pressure depends on the entropy of fusion per unit volume of water, rather than ice.) Since $(S_L - S_C)/v_L \approx 1.3$ MPa/°C for water, an undercooling of just 3°C generates a negative pressure of $P_L \approx -3.9$ MPa. Buil and Aguirre-Puente[13] directly measured the pressure exerted by ice against a membrane and quantitatively verified Eq. 12. It is this pressure that is responsible for the shrinkage of the air-entrained sample shown in Fig. 1.

From Eqs. 10 and 11, we find that the curvature at the nose of the crystal in Fig. 4 is

$$\gamma_{CL} \kappa_{CL}^E = -\left(\frac{S_L - S_C}{v_L}\right)(T_m - T) \tag{13}$$

At any point on the crystal/liquid interface where the curvature differs from κ_{CL}^E, there is a pressure on the wall that satisfies Eq. 6; at the shoulder, S, where the interface is flat, $P_A = \gamma_{CL} \kappa_{CL}^E$. As the temperature decreases, the curvature at the nose increases and permits penetration of the pore when

$$\kappa_{CL}^E = \frac{2}{r_p - \delta} \tag{14}$$

At this point, according to Eqs. 12 and 13, the pore pressure is

$$P_L - P_e = -\gamma_{CL}\kappa_{CL}^E = \frac{2\gamma_{CL}}{r_p - \delta} \qquad (15)$$

Now, according to Eq. 6, as long as the pore is convex ($\kappa_{CL}^S \geq 0$), the maximum pressure the crystal can exert on the pore wall is no greater than $\gamma_{CL}\kappa_{CL}^E$; this is likely to be true even at constrictions in the pore, as shown in Fig. 3(b). Therefore, the pressure exerted on the wall when the crystal enters is not larger than the suction given by Eq. 15, and the net stress on the wall is compressive, so long as the pore pressure remains in equilibrium with macroscopic ice. Unfortunately, such equilibrium may not be preserved. If the crystal in Fig. 4(a) enters the pore, the volume increase as water freezes will push water toward the liquid/vapor meniscus, tending to flatten it; this will raise the vapor pressure over the meniscus, and the excess water will evaporate. However, in view of the rapid crystallization rate of ice,[17,18] the rate of evaporation may not be able to keep pace. In that case, once the ice penetrates the pores, the body will expand.

Ice will first enter the largest pores touching the air void, but these do not generally percolate throughout the body. At any given temperature, the ice will advance through the pore network until the perimeters of the ice crystals are all bounded by pores too small to satisfy Eq. 8. This situation is illustrated in Fig. 5(b). During an isothermal hold, the stress exerted on the pore walls by the ice should remain constant, but Powers and Helmuth[4] observed that air-entrained paste continued to contract for several minutes after cooling was arrested. This probably means that the crystal growth was so fast during cooling that the pore pressure did not remain in equilibrium with the macroscopic ice; during the hold, P_L decreased toward the value given by Eq. 12, resulting in a small contraction. The explanation for the contraction offered by Helmuth[9] was that crystals in the mesopores would melt because they are subjected to greater pressure than macroscopic crystal in air voids. However, the preceding analysis indicates that, owing to the negative pore pressure, the large and small ice crystals are in equilibrium regardless of their size.

As T drops and the ice enters smaller pores, it may eventually be able to penetrate pores with radii smaller than the breakthrough radius, r_{BT}.[6] This is the largest entrance into the network of pores that percolates throughout the body; it corresponds roughly to the peak of the pore size distribution found

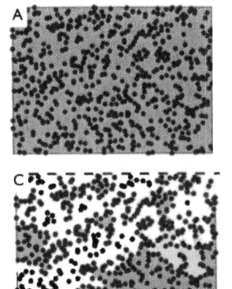

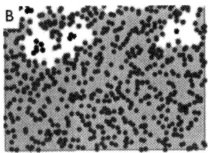

Figure 5. (a) Cross section of a saturated porous body; black dots represent the solid phase; shading is liquid. (b) Ice has invaded the larger pores that connect to the surface, but the leading edge of the freezing front is blocked by pores too small for the menisci to enter. (c) The temperature has fallen below T_{BT}, so the ice has entered the percolating network of pores that span the body.

from nitrogen desorption,[19] mercury intrusion,[20] or thermoporometry, and is the characteristic size that controls the liquid permeability.[21] Once the temperature is low enough to satisfy Eq. 8 with $r_p = r_{BT}$, then the ice can advance arbitrarily far through the pore network, as in Fig. 5(c). Therefore, the freezing front can pass through the body when

$$\kappa_{CL} = \frac{2}{r_{BT} - \delta} \tag{16}$$

which happens when T drops to the breakthrough temperature, T_{BT}; from Eq. 2,

$$T_{BT} = T_m - \frac{2\gamma_{CL}}{(r_{BT} - \delta)\Delta S_{fv}} \tag{17}$$

Thus, the propagation rate of ice through the pores is zero until $T < T_{BT}$, then it increases at lower temperatures. However, the growth of ice releases heat of fusion, which causes the temperature of the surroundings to rise. In bulk water, the temperature rises to 0°C, and the rate of growth becomes controlled by the rate at which heat diffuses away from the crystal.[22] Similarly, the heat of fusion causes the temperature in a porous medium to rise to T_{BT}. This was demonstrated in an experiment by Helmuth,[18] where he showed that the temperature of cement paste rises to a fixed value during freezing, regardless of the exterior temperature; the internal temperature depends on the pore structure of the sample. (Unfortunately, he did not measure the pore size distribution of his samples so as to provide a quantitative test of this concept; however, the temperature was lower in samples with lower water/cement ratio, and consequently smaller pores.) The ice thus propagates through the percolating network of pores at a temperature approximately equal to T_{BT}.

When the ice percolates through the capillary network, (assuming that the pores are roughly cylindrical) the crystallization pressure is given by Eq. 7:

$$\sigma_\theta \approx \frac{\gamma_{CL}}{r_{BT} - \delta} \qquad (18)$$

However, according to Eqs. 10, 11, and 16, the pressure in the liquid is

$$P_L - P = -\frac{2\gamma_{CL}}{r_{BT} - \delta} \qquad (19)$$

Again, as long as equilibrium is preserved, the negative pore pressure offsets the crystallization pressure, and the net stress in the solid network is compressive. The body will expand if the growing ice falls out of equilibrium with the macroscopic ice, which may happen in concrete even at the slow cooling rates (~5–6°C/h) occurring in nature.

Experimental Studies of Dilatation during Freezing

Litvan[2] measured the dimensional change in porous Vycor glass that was saturated with water and frozen. As shown in Fig. 6, the initial freezing (which was observed to occur on the outer surface of the sample) causes contraction, and the body continues to contract as the temperature drops to about –18°C. The negative pressure in the pores causes a maximum strain

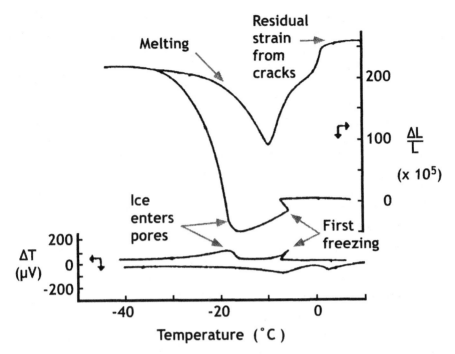

Figure 6. Data from Ref. 2 showing linear strain (right axis) and temperature change (left axis) in water-saturated porous Vycor glass. The temperature rises when nucleation of ice occurs at the exterior of the sample (first freezing). As the temperature decreases, the pore pressure becomes increasingly negative, according to Eq. 12, and the body contracts. At about −18°C, Eq. 14 is satisfied and the ice invades the pores, resulting in expansion; the crystallization pressure causes cracks that produce a residual expansion.

of about 6×10^{-4}, or $3.3 \times 10^{-5}/°C$. According to Biot's constitutive equation for a porous elastic solid,[23,24] the linear strain, ε_x, caused by the pore pressure is

$$\varepsilon_x = \frac{P_L}{3K_p}\left(1 - \frac{K_p}{K_S}\right) \tag{20}$$

where K_p is the bulk modulus of the porous body and K_S is the bulk modulus of the solid phase itself. For Vycor, $K_p \approx 7.8$ GPa and the modulus ratio is estimated to be $K_p/K_S \approx 0.42$,[25] so the pressure causing the contraction in Litvan's experiment is found from Eq. 20 to be $P_L = -(3.3 \times 10^{-5})(3)(7.8 \times$

$10^9)/0.58 = -1.3$ MPa/°C, which is the value expected for water, according to Eq. 12. As the temperature drops further, the sample begins to expand, apparently owing to invasion of the pores by ice; according to Eq. 8, an invasion temperature of −18°C corresponds to a pore radius of ~4 nm, which is reasonable for Vycor. Since the pore size is quite uniform in this material, the pore space is invaded over a narrow range of temperature and crystallization pressure is exerted on the entire pore volume, leading to such a large expansion that cracking occurs. The fact that the sample expands indicates that the pore liquid does not remain in equilibrium with the solid ice on the surface. There may also be damage from pockets of water that are entirely surrounded by ice[10,26]; as ice invades the trapped pores, the volume change can generate pressure of 13.5 MPa/°C.[14]

Negative pore pressure also accounts for the contraction reported by Powers and Helmuth in air-entrained samples,[4] where macroscopic ice crystals could form in the voids. Helmuth[9] reported that air-entrained samples with water/cement ratios of 0.56–0.64 contracted at a rate of 35×10^{-6}/°C upon cooling below 0°C; assuming that the thermal expansion coefficient of the solid paste is about 10×10^{-6}/°C,[27] the contraction attributable to pore pressure is about 25×10^{-6}/°C. The net pressure causing contraction in Helmuth's experiment is $P = P_L + \phi P_A$, where ϕ is the volume fraction of the body penetrated by ice. (Note that all of the contiguous water can be drawn into tension by the ice, even before it enters the pores; however, the pressure P_A is exerted only over the volume fraction of pores penetrated by the freezing front.) Measurements in our lab[28] on paste with w/c = 0.6 indicate that Young's modulus is $E_p \approx 12$ GPa; estimating Poisson's ratio to be $\nu_p \approx 0.20$, the bulk modulus is $K_p = E_p/[3(1-2\nu_p)] \approx 6.7$ GPa. The results in Ref. 28 indicate that $K_p/K_S \approx \rho^2$, where ρ is the volume fraction of solids, and for this paste $\rho \approx 0.5$. Using these values in Eq. 20, we find that Helmuth's experiments imply that the net pressure causing the contraction was $P = -(25 \times 10^{-6})(3)(6.7 \times 10^9)/0.75 = -0.7$ MPa/°C. Since Eq. 12 indicates that $P_L = -1.3$ MPa/°C, and part of that pressure (depending on the shapes of the pores and the volume fraction of the pores containing crystals) would be offset by P_A, Helmuth's results are in reasonable quantitative agreement with Eq. 20.

Suppose that the first nuclei appear in the mesopores, rather than in large voids or at the free surface. In a paste prepared with a high w/c ratio, as in the samples used by Powers and Helmuth,[4] there will be little self-desiccation, so the pore pressure will be low ($\kappa_{LV} \approx 0$); therefore, Eq. 9 indicates that

$$\gamma_{CL} \kappa_{CL}^{E} = \left(\frac{S_L - S_C}{v_C} \right)(T_m - T) \tag{21}$$

In cylindrical pores, P_A will be ~0.5–1 times this value (depending on the shape of the pores). This would appear to account for the expansion of the sample in Fig. 1 that was not air-entrained. If that expansion had been caused by hydraulic pressure, it would have relaxed away during the isothermal hold, rather than remaining constant, so it was most likely caused by crystallization pressure (as was later suggested by Helmuth[9]). However, since the volume fraction of the pore water that freezes is small,[9,29] the crystallization pressure is too small to account for the observed expansion. There may be stress from freezing of entrapped pockets of water,[26] or the invasion of the ice may provoke progressive microcracking.

If the first ice to nucleate occurs in the mesopores, then the body will expand; this is seen in paste with or without air entrainment.[9] However, as soon as ice nucleates at a free surface, the pore pressure will begin to drop toward the negative value given by Eq. 15; this will allow the curvature of the ice crystals to increase from the value given by Eq. 21 to that given by Eq. 13, which is about 10% larger. The volume fraction of the pores occupied by ice will therefore increase, but the change in the crystallization pressure will be more than offset by the reduction in pore pressure, so the body will contract as long as equilibrium is preserved.

Role of Air Voids

Our analysis indicates that damage can result from crystallization pressure, unless macroscopic crystals nucleate in air voids. Given that (1) expansion is seen when ice first nucleates,[4] (2) the magnitude of the initial expansion is greater when the cooling rate is slower (so that a greater amount of ice forms in a given temperature interval),[9] and (3) nucleation occurs near –8°C in many of the tests reported in the literature (e.g., Refs. 18, 30–32), it appears that the first nucleation events occur in the mesopores. Therefore, it would be useful to promote nucleation in the voids at a temperature close to 0°C. Nucleating agents have been applied to the surface of concrete to favor nucleation,[29] and it has been proposed to introduce them into the air voids.[33] However, it may be that the primary function of air-entraining agents is to promote nucleation. Chatterji[34] notes that some air-entraining agents (AEA) are more effective than others, and attributes their efficiency to their influence on the adhesion of ice to the interior surface of the void (which, he argues, makes the ice more efficient at extracting water from the pores). Vinsol resin is an AEA that is said to be particularly protective,[34] and it is also one that forms a distinct shell of calcium-rich deposits at the surface of the voids.[35] It may be that the precipitates that form on the layer of AEA on the surface of the air void favor nucleation of ice, in which case their differing effectiveness may be related to the evident difference in the way they affect the structure of the adjacent paste. In the microscopy study by Corr et al.,[36] hemispherical crystals of ice were observed to nucleate on the surface of air voids. The role of AEA in nucleating ice deserves further study.

As suggested above, it would also be beneficial to have relatively small pores surrounding air voids, to allow the macroscopic ice to extract water from the mesopores without invading them. Indeed, the interaction of AEA with hydrating paste[35,36] may also refine the pore structure adjacent to the void, and thereby facilitate extraction of water from the mesopores.

The classical interpretation of the role of air voids is that they serve as sinks for water pushed ahead of the freezing front.[1] This idea was later cast into doubt[4,9] because it was demonstrated that the expansion during freezing did not increase with the cooling rate, as the theory predicts. Fagerlund[37] showed that frost damage occurs rapidly when a certain critical degree of saturation, S_{CR}, is exceeded; for air-entrained concrete, S_{CR} corresponded to a condition where >40% of the air voids were full of water. (That is, all of the voids will contain some water, but the smaller ones will be completely filled.[38]) Clearly, if ice nucleates in a full air void, the sudden volume increase creates a pressure pulse that cannot be accommodated by flow into the surrounding paste. For example, using the crystallization rates measured by Hillig and Turnbull,[17] at −1°C the growth rate is 1600 μm/s, so all the water in a 200 μm void would crystallize in ~0.12 s. If the surrounding paste were 50% porous, the velocity of the water entering the pores would be J ≈ 3200 μm/s. The resulting pore pressure can be estimated using Darcy's law,[39]

$$ J = -\frac{k}{\eta} \nabla P_L \tag{22} $$

where k is the permeability of the paste and η is the viscosity of the liquid.

Supposing optimistically that the pressure could be relieved by flow to an air void at atmospheric pressure at a distance $L = 200$ μm away, the pressure gradient would be ~P_L/L, so $P_L \approx J \times \eta \times L/k$. Given $\eta \approx 0.001$ Pa·s, we find $P_L \approx 6.4 \times 10^{-10}/k$, so even for a relatively permeable paste with $k \approx 10^{-19}$ m$^2 \approx 10^{-12}$ m/s,[28] the pressure would be in the gigapascal range. Of course, the actual pressure is limited by the Clausius-Clapeyron condition, so it cannot exceed 13.5 MPa/°C of undercooling[14]; however, that substantially exceeds the tensile strength of concrete, so cracking would be immediate. Thus, the damage mechanism in Fagerlund's experiments was not classical hydraulic pressure resulting from crystallization of ice in capillary pores; rather, it was from the explosion of water-filled air voids. For any lower degree of saturation, the concrete was found to contract during cooling, as expected when crystallization in larger pores creates suction in the mesopores.

Salt Scaling

When a pool of liquid freezes on a concrete surface, superficial damage called salt scaling often occurs. The phenomenon has been described in several excellent reviews.[40–42] The principal observations are as follows:

1. The damage is worst when the water contains a moderate amount (the so-called pessimum concentration) of solute.[42–44]

2. The pessimum concentration is nearly independent of the nature of the solute (e.g., salts, alcohol, and urea show similar behavior[44]).

3. The damage consists of small flakes of material removed from the surface.[45]

4. No scaling occurs without free liquid on the surface of the sample (i.e., saturated surface-dry samples do not scale).[44,46]

5. Damage is worse when the minimum temperature in the cycle is lower.[42,46,47]

6. The salt concentration in the exterior liquid is more important than that of the pore liquid.[40,42]

7. Entrained air reduces the damage.[40,44]

Many theories have been offered to account for some of these facts, but none of the proposed explanations of the mechanism is entirely satisfactory. For example, it has been suggested that the damage results from thermal shock when salt melts ice on the surface of a road or sidewalk, but field measurements do not support the occurrence of destructive temperature changes.[42] Moreover, lab experiments do not involve the addition of salt to existing ice; instead, the salt is dissolved in the pool of free liquid on top of the sample. Damage has been attributed to crystallization just below the surface, where saturation is made possible by the free water at the outer surface[42]; however, testing shows damage after tens or hundreds of freezing cycles, by which time the saturation should extend well beyond the superficial zone.[45] Moreover, the saturation should also be present when the pool contains pure water, but that does not cause scaling.[46] Superficial damage has been blamed on over-finishing or bleeding, which alter the surface structure, and it has been demonstrated[49] that the depth of scaling matches the depth of carbonation (in blast furnace slag cement). These factors may account for the depth of the damage, but do not explain what causes it. It has been suggested that the damage results from precipitation of salt, but solid sodium chloride does not form above −22°C, which is below the temperature range of most scaling tests. It has been argued that increased saturation of the surface layer occurs when salt is present in the pore liquid,[32,42] but that effect would increase monotonically with salt concentration, not show a pessimum. Finally, it has been proposed that osmotic pressure drives water from the pores toward the free liquid[10]; again, this effect would increase monotonically with salt concentration, so it cannot account for the pessimum. We will show that dilatation does occur when cement paste is exposed to salt, but that it apparently results from a combination of crystallization pressure and ion exchange; there is no evidence of a significant osmotic pressure.

This chapter examines mechanisms that have not been previously considered: (1) the negative pressure created in the pore liquid by surface crystallization that creates differential strain, resulting in tensile stress at the surface; (2) mechanical bonding of the ice to surface irregularities on the concrete that allows the ice to exert stresses during cooling, owing to the huge mismatch in thermal expansion coefficient between ice and concrete; and (3) dilatation caused by exposure of paste to salt.

Pore Pressure

Figure 7 shows the experiment schematically. A pool of liquid on top of a block of paste or concrete is frozen and it creates suction in the pore liquid, according to Eq. 12. This phenomenon was shown to cause contraction in Figs. 1 and 6. If the sample is large enough that the pore pressure equilibrates slowly, then the shrinkage will be localized near the surface, and the

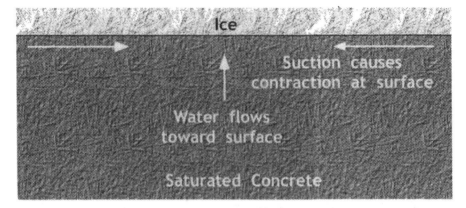

Figure 7. When ice forms on the surface and before it invades the pores, liquid is drawn from the pores to freeze onto the macroscopic crystal. The resulting suction in the pores causes contraction of the surface region. If the pressure gradient is steep, the differential strain can cause significant tension in the surface region.

resulting differential strain could cause damage. For example, using the property values cited earlier for the paste in Fig. 1, if Young's modulus is 12 GPa and the tensile strength is 3 MPa, then the failure strain is 2.5×10^{-4}. A strain of that magnitude is seen to occur upon cooling to about $-7°C$. As noted earlier, it is common in lab tests for nucleation to occur at temperatures in that range, so it would be possible to have a sudden freezing event that created pore pressure at the surface high enough to cause cracking.

The magnitude of the stresses and strains can be calculated by using Biot's elastic constitutive equation for a saturated porous body[50] and assuming that liquid transport obeys Darcy's law. The same formalism has been successfully applied to describe the response of saturated cement paste to bending[24] and thermal stresses.[51] The analysis[52] indicates that under realistic conditions (L = 10 cm, $dT/dt = -5°C/h$) the stress could exceed 1 MPa before the temperature drops to T_{BT}. That is enough to cause growth of fatigue cracks over many freeze/thaw cycles. The characteristic time for relaxation of the pressure gradient can be on the order of hours for a material with a low permeability[27]; the stress will eventually dissipate as the pressure equilibrates within the body and the strain becomes uniform.

To investigate the pressure gradient, we built an apparatus that is shown schematically in Fig. 8. A thin plate of cement paste is supported at the ends (on low-expansion Invar metal posts) and covered with a pool of liquid (within a dam created using vacuum grease), then it is cooled to the freezing point. If a pressure gradient were created as the sample was cooled, the

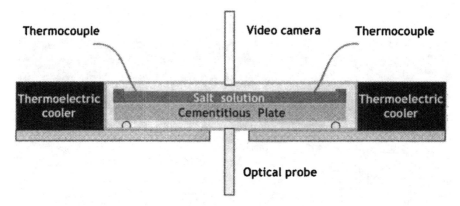

Figure 8. Schematic of apparatus to measure the pressure gradient in pore liquid. A plate of cement paste on Invar supports is covered with a pool of liquid; the system is inside a copper box, which is inside an insulated enclosure (not shown). The temperature is reduced by Peltier coolers until the water freezes. Any resulting pressure gradient causes bending of the sample, which is detected by the optical probe.

deflection (warping) would be detected by the optical probe. For example, Fig. 9 shows the temperature and deflection measured using a plate (0.3 × 2.1 × 10 cm) of cement paste covered with a pool 0.3 cm deep containing 3 wt% NaCl in water ($T_m = -1.8°C$).[52] The temperature is decreased in steps of about 5°C from room temperature, and each time T decreases, the deflection increases and then relaxes. The reason for this is as follows. When T drops, the pore liquid contracts more than the solid phase, so the liquid withdraws into the pores, creating negatively curved menisci on the bottom surface. On the top, the pool of liquid prevents the formation of negative pore pressure, because the free liquid is sucked into the pores. As a result, there is a gradient in pore pressure from atmospheric pressure near the top surface to negative pressure on the bottom, and this causes the plate to deflect upward (becoming concave downward). When T is held constant, liquid from the pool is drawn into the pores to relieve the gradient, and the plate flattens out. The same plate of cement was subjected to three-point bending,[52] which also creates an internal pressure gradient in the pore liquid[53]; the time required to relax that gradient was comparable to the time to relax the deflections following temperature changes. Nucleation of ice occurred at −3.8°C (~230 min), as revealed by the jump in temperature (to T_m) as the heat of fusion was released. The heat effect, together with the reinforcing effect of the ice layer, apparently obscures the deflection expected from the pressure gradient produced by freezing. However, two more

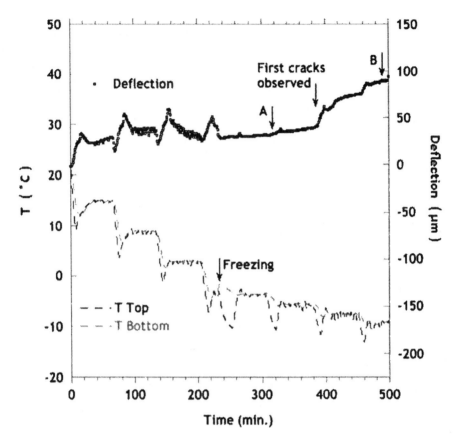

Figure 9. Surface temperature and deflection of plate of cement paste as function of time; the plate is 0.3 cm thick and 10 cm long, and is covered with ~0.3 cm of a solution containing 3 wt% NaCl. Freezing occurs at about 230 min, when the temperature reaches −3.8°C; the first cracks in the ice are seen at ~400 min. Points A and B correspond to the photos shown in Fig. 11.

important effects were revealed by the experiment, as detailed below.

Salt-Induced Expansion

Underlying the deflections caused by the temperature steps in Fig. 9 is a slow rise in the baseline. This was found to result from expansion of the surface of the paste upon exposure to the salt solution. Figure 10 shows the deflection of a similar plate of cement paste at room temperature when an aqueous solution of NaCl is put on its surface. The plate deflects upward

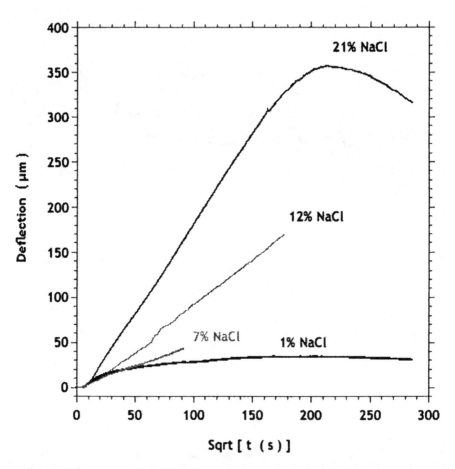

Figure 10. Deflection of plate (0.3 × 2.1 × 10 cm) of cement paste at room temperature when aqueous solution with indicated concentration (wt%) of NaCl is poured on top.[52]

(indicating expansion of the paste) for about 2 days, then begins to flatten out; the decrease in curvature is expected once material below the midplane of the plate begins to expand. It might be expected that there would be an osmotic effect, as water from the pores diffuses into the pool of brine, but movement of water out of the pores would cause warping in the opposite direction from what is observed. Moreover, if water diffused from the plate, the resulting suction would draw liquid from the pool into the pores, and the relaxation time would be a matter of minutes, not tens of hours. In fact, it has been observed[54] that various types of cement expand upon exposure to salts. Alkali adsorb reversibly on C-S-H, apparently binding to deprotonated silanols[55,56]; chloride adsorbs, largely irreversibly[57]; and simulations indicate that the primary adsorption site is the surface of calcium hydroxide crystals.[58] Recent XRD analysis[52] indicates that the expansion is accompanied by the formation of Friedel's salt, so crystallization pressure may contribute to the observed strain.

We are currently investigating the kinetics of expansion of paste as a function of salt concentration and type. What is evident, however, is that the expansions are very large. This may be highly relevant to scaling, because as the NaCl solution freezes, the remaining liquid becomes highly concentrated in salt, approaching the eutectic concentration of 21%. Therefore, during each cooling cycle, the surface of the paste is exposed to brine at that high salt concentration and will expand; then, during the heating cycle the ice will melt and the liquid will rehomogenize at its original concentration, causing the surface of the paste to contract. Regardless of the initial concentration of salt (as long as it is >0), when the temperature reaches –20°C, the remaining brine will contain ~20 wt% NaCl. If the initial salt concentration is low, there is a very large variation in brine concentration upon melting, but if the initial concentration is high, the variation is correspondingly lower. This effect could contribute to the existence of a pessimum concentration: for salts with low, but nonzero, initial concentrations, every freezing cycle produces a cycle of strong compressive stress at the surface. Preliminary dilatometry experiments using a paste with w/c = 0.45 in 21 wt% NaCl indicate a maximum strain of $\varepsilon_x \approx 6 \times 10^{-4}$. If this were confined to a thin layer on the surface of a block of concrete (as in a typical scaling test), the compressive stress would be $\sigma_x = E\varepsilon_x/(1-v)$; given $E \approx 15$ GPa and $v \approx 0.2$, the compressive stress would be ~11 MPa. Over the course of many cycles, a stress of this magnitude could cause fatigue damage to the surface, particularly where thin layers of paste overlie grains of fine aggregate.

Bimaterial Effect

The experiments revealed another effect that may be an important factor in scaling damage: a layer of pure ice causes severe warping of the sample owing to the mismatch in thermal expansion in the bimaterial composite of ice and paste. We know that this is the cause of the warping, because the deflection does not relax rapidly during an isothermal hold, as the pressure gradient does, and the deflection is found[52] to be in agreement with the deflection calculated from Timoshenko's equation for a bimaterial strip.[59]

The stress produced from this mechanism can amount to a few MPa, and increases as the temperature decreases. Damage could result from shearing of asperities from the surface at points where the ice and cement paste are interlocked. However, a more likely mechanism is that the frozen layer cracks, and the crack continues a short distance into the concrete, resulting in superficial damage. A mechanism of this kind is used to produce a decorative surface on glass[60]: a layer of epoxy is spread on a plate of glass, which is then cooled until the thermal expansion mismatch causes the epoxy to crack; the cracks penetrate into the glass and remove a thin layer, resulting in a scalloped surface. The lower the temperature to which the body is cooled, the greater the probability that the layer will crack, so this mechanism is consistent with experiments that indicate that scaling damage is worse when the minimum temperature of cooling is lower.[42,61]

Experimental observations and a viscoelastic analysis indicate that pure water ice does not crack from thermal expansion mismatch with cement paste.[62] This may explain why scaling is not serious when pure water is used in the surface layer. However, when any amount of salt is present, then the ice contains pockets or channels of brine that constitute mechanical flaws and encourage fracture of the layer. This could account for the pessimum: pure ice doesn't crack in the temperature range used in scaling tests (typically down to $-20°C$), and high salt concentrations yield ice with so much brine that they cannot impose significant stress at $-20°C$; however, intermediate salt contents permit development of substantial stresses and contain flaws that will promote cracking when the stress becomes high. This will lead to damage from low to moderate salt concentrations, but less damage at high concentrations.

The occurrence of cracking in a layer of ice made from a 3% solution of NaCl is demonstrated by the photos in Fig. 11, which were obtained by introducing a video camera lens through the top of the box, as indicated in Fig. 8. The upper photo corresponds to point A in Fig. 9, shortly after freezing has begun. Stress from the expansion mismatch offsets the stress from salt-induced expansion, so little deflection is occurring; no cracks are present in the ice. The first cracks are observed at about 400 min, when the temperature reaches $-7.5°C$; weakening of the ice by the cracks results in upward deflection because of the salt-induced swelling. Fig. 11(b) shows numerous cracks in the ice at the end of the experiment (490 min); the largest ones are indicated by arrows. The fact that there is no liquid on the top surface indicates that freezing proceeded from the air/water interface toward the paste, pushing the brine ahead of the freezing front. However, the occurrence of cracks proves that the ice was strongly adhering to the

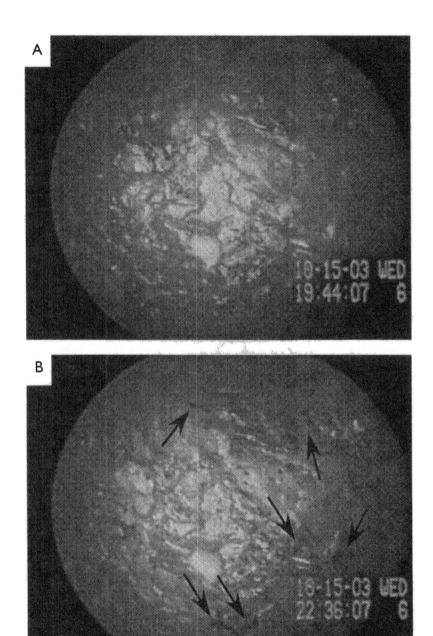

Figure 11. Photos of an ice sheet on the surface of a plate of cement paste before (a) and after (b) cracking of the ice sheet; major cracks are indicated by the arrows. The top photo corresponds to point A in Fig. 9 and the lower photo to point B. The width of the field of view is ~0.6 cm.

paste; therefore, the surface of the paste is exposed to stress from the bimaterial effect and is also in contact with the unfrozen brine.

The bimaterial effect accounts for all of the phenomenology of scaling.[63] Damage is seen only when there is free liquid on top, because a certain thickness of ice is needed to generate the stress. The pessimum results from the mechanical properties of brine: pure ice does not crack, but impure water freezes into ice containing brine pockets that constitute mechanical flaws. This explains why damage diminishes at high solute levels, where the ice is too weak to exert appreciable stress, and why the nature of the solute does not matter. Cracking of the ice creates tensile stresses where the ice joins the surface of the paste, and a fracture mechanics analysis reveals that damage is expected to result,[64] especially in weak surfaces. On the other hand, salt-induced swelling causes cycles of compressive stress that could cause buckling of thin layers of paste over aggregate particles. This phenomenon could also contribute to the pessimum, because the intensity of the composition swings on freezing and thawing is maximal when the original solution has a small amount of salt. The relative importance of these two mechanisms is currently under study in our lab.

The beneficial effect of air entrainment is more difficult to explain on the basis of either of these mechanisms. It may be that air entrainment, by reducing bleeding, leads to greater strength in the superficial layer of paste and thereby reduces damage. However, this remains to be proven, so the effectiveness of air entrainment presents an important challenge to the bimaterial or salt-swelling mechanism.

Conclusions

Frost damage is a major cause of deterioration of the infrastructure, and has therefore been the subject of intensive study for decades. In spite of the level of effort devoted to the problem, there is still disagreement about the fundamental mechanisms responsible for the damage. The preponderance of evidence indicates that the primary source of stress during freezing is crystallization pressure, not hydraulic pressure. The role of air voids is to provide sites for nucleation of macroscopic ice crystals that can grow without exerting pressure on the solid phase. When such crystals form, they draw liquid from the mesopores and cause the body to contract; in the absence of air voids, crystals nucleate in the small pores where they exert stress on the pore walls, leading to expansion and cracking. In porous glass,[2] it has been demonstrated that expansion occurs even when the pores contain a normal liquid that is less dense than its crystal phase; in such cases, there is no positive hydraulic pressure, so the expansion can be attributed only to crystallization pressure.

When crystals nucleate at free surfaces, the contraction produced by the negative pore pressure can be calculated using classical thermodynamics. We have shown that shrinkages measured on porous glass and cement paste can be quantitatively explained in this way. If air entrainment is not sufficient, nucleation of ice in the mesopores produces local stress concentrations that can lead to cracking, particularly if saturated zones are entirely surrounded by ice. We suggest that the best defense against such damage is to encourage nucleation in the air voids, perhaps by introducing nucleating agents, and to have relatively small pores surrounding the voids to inhibit penetration of the pores by the ice.

The mechanism of salt scaling is still unclear, and no mechanism has previously been proposed that accounts for all of the experimental observations. We propose two ideas that account for most of facts. First, we observe that a strongly adherent ice layer can generate high stresses owing to its thermal expansion mismatch with cement paste; small amounts of entrapped brine cause the layer to crack and this may damage the underlying paste. Second, we find that exposure to salt solutions causes substantial expansion of cement paste; variations in the brine concentration during freeze/thaw cycles could lead to compressive failure at weak points on the surface of the paste.

Acknowledgment

This work was supported by National Science Foundation Grant CMS-0200440.

Appendix: Equilibrium for Small Crystals

Gibbs[65] derived an expression for the chemical potentials of a small crystal and the surrounding liquid and showed that the two were not equal. The following treatment uses his procedure, but extends his derivation to allow for a multicomponent solid with a density different from that of the liquid and with pressure applied only to the solid phase. The pressure represents the constraint imposed by an obstacle, such as a pore wall, that blocks the growth of a crystal. Gibbs's approach is to consider the transfer of a small amount of material between a solid body and a large surrounding bath of liquid. He accounts for the energy in the increment of solid removed (or added) and the increment of liquid formed from it (or from which it forms);

all changes in energy and composition are provided by the bath, so that the energy of the entire system remains constant.

Notation

E = energy

S = entropy

P = pressure

T = absolute temperature

V = volume

v = molar volume

$\bar{V}_i$ = partial molar volume of component i

n_i = moles of component i

R_1, R_2 = principal radii of curvature of the solid surface

κ_{CL} = $1/R_1 + 1/R_2 = dA/dV$ = mean curvature of the solid surface

$\mu_i = (\delta E/\delta n)_{S,V}$ = chemical potential of component i

Γ_1 = superficial concentration of component i

γ_{CL} = surface energy of the isotropic crystal/liquid interface

f_{CL} = surface stress in the crystal/liquid interface

δA = element of area on the crystal/liquid interface

δN = incremental displacement normal to δA (negative for dissolution or melting)

Subscripts

C = crystal

L = liquid

I = interface

Consider a multicomponent isotropic solid, containing N_C components ($i = 1, 2, \ldots N_C$) immersed in a liquid containing $N > N_C$ components. We will find the energy change associated with dissolution or melting of a small amount of solid; the components released by the solid become part of the large surrounding bath of liquid, producing a negligible change in composition.

The energy of the crystal is

$$E_C = TS_C - P_C V_C + \sum_{i=1}^{N_C} n_{Ci} \mu_{Ci} \qquad (23)$$

and the energy of the liquid is

$$E_L = TS_L - P_L V_L + \sum_{i=1}^{N} n_{Li} \mu_{Li} \qquad (24)$$

The total number of moles of crystal, n_C, is

$$n_C = \sum_{i=1}^{N_C} n_{Ci} \qquad (25)$$

If the molar volume of crystal is v_C, then the total volume of crystal is

$$V_C = \sum_{i=1}^{N_C} n_{Ci} \bar{V}_{Ci} = n_C v_C \qquad (26)$$

Owing to the difference in partial molar volumes of the components in the liquid and solid state, the molar volume of the liquid created when the crystal dissolves is

$$v_{CL} = \frac{1}{n_C} \sum_{i=1}^{N_C} n_{Ci} \bar{V}_{Li} \qquad (27)$$

The energy per unit area of the crystal/liquid interface is

$$E_I = TS_I + \gamma_{CL} + \sum_{i=1}^{N} \Gamma_i \mu_{Li} \qquad (28)$$

where γ_{CL} is the energy needed to create new area by adding material to the interface; in general, this differs from the surface stress, f_{CL}, needed to increase the surface area by stretching it.[65] The sum in Eq. 28 begins at $i = 2$, because the location of the interface is chosen so as to make the concentration of component 1 equal to zero.

As the crystal transforms, it changes in volume by $\delta A \delta N$ (which is negative for dissolution or melting) and the corresponding change in energy of the crystal is

$$\Delta E_C = \int \frac{E_C}{V_C} \delta A \delta N \tag{29}$$

The volume of liquid produced during the change of state is (v_{CL}/v_C) $\delta A \delta N$. The infinitesimal amount of solid does not change the composition of the liquid, so the energy per unit volume of liquid is unaffected. Therefore, the change in energy of the liquid resulting from the transformation is

$$\Delta E_L = -\int \left(\frac{E_C}{V_L}\right)\left(\frac{v_{CL}}{v_C}\right) \delta A \delta N \tag{30}$$

The energy per unit area of the crystal/liquid interface is E_1, and the change in area of the solid is $dA = \kappa_{CL} dV = \kappa_{CL}\delta A \delta N$, so the change in interfacial energy is

$$\Delta E_1 = \int E_1 \kappa_{CL} \delta A \delta N \tag{31}$$

The change in entropy of the crystal and adjacent liquid is

$$\Delta S = \int \left[\frac{S_C}{V_C} - \left(\frac{S_L}{V_L}\right)\left(\frac{v_{CL}}{v_C}\right) + S_1 \kappa_{CL}\right] \delta A \delta N \tag{32}$$

and the change in chemical components is

$$\Delta n = \int \left[\sum_{i=1}^{N_C} \frac{n_{Ci}}{V_C} - \sum_{i=1}^{N_C} \frac{n_{Li}}{V_L}\left(\frac{v_{CL}}{v_C}\right) + \sum_{i=2}^{N} \Gamma_i \kappa_{CL}\right] \delta A \delta N \tag{33}$$

All of the changes in energy and chemical components produced by the transformation are assumed to be provided by the surrounding liquid bath. The energy change of the bath is obtained by multiplying the entropy change (Eq. 32) by $-T$ and the change in moles of each component (Eq. 33), by its chemical potential. In addition, we have to take account of the work done by the crystal against the applied pressure. We want to allow for crystallization pressure, which is the pressure exerted on a growing crystal

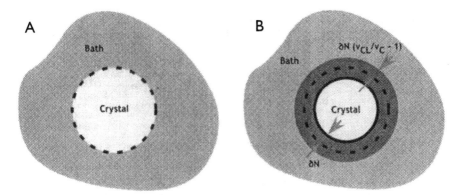

Figure 12. (a) A crystal is at equilibrium in a large bath; the dashed line indicates the location of the crystal/liquid interface. (b) The interface is displaced a distance δN by dissolution or melting; the liquid created by the phase change (indicated by dark shading) occupies a volume that differs by a factor of v_{CL}/v_C from that of the solid from which it was produced.

by an obstacle, such as a pore wall, so we include a pressure, P_A, applied on the crystal (but not on the liquid). The total pressure on the solid is thus

$$P_C = P_L + P_A + f_{CL} \kappa_{CL} \tag{34}$$

where f_{CL} is the surface stress. As indicated in Fig. 12(a), we begin with a crystal immersed in a large bath; in Fig. 12(b), the radius of the crystal has contracted by δN, and a quantity of liquid has formed (shown by darker shading) that occupies a volume different from that of the solid that has transformed. As the crystal shrinks, the pressure on the crystal does work equal to $P_A \delta N \delta A$, which goes into the bath. At the same time, the newly created liquid pushes the boundary of the bath back by $\delta N \delta A (v_{CL}/v_L - 1)$, so it does work against the liquid pressure, P_L, which must be removed from the bath. Therefore, the total energy change of the bath is

$$\Delta E_{\text{bath}} = -T \int \left[\frac{S_C}{V_C} - \left(\frac{S_L}{V_L} \right) \left(\frac{v_{CL}}{v_C} \right) + S_I \kappa \right] \delta A \delta N$$

$$- \int \left\{ \left[\sum_{i=1}^{N_C} \frac{n_{Ci}}{V_C} - \sum_{i=1}^{N_C} \frac{n_{Li}}{V_L} \left(\frac{v_{CL}}{v_C} \right) \right] \mu_{Li} + \sum_{i=2}^{N} \Gamma_i \mu_{Li} \kappa_{CL} \right\} \delta A \delta N$$

$$+ \int \left[P_A - P_L \left(\frac{v_{CL}}{v_C} - 1 \right) \right] \delta A \delta N \tag{35}$$

The total energy change of the system is

$$\Delta E = \Delta E_C + \Delta E_L + \Delta E_I + \Delta E_{\text{bath}} \tag{36}$$

When the terms are assembled, the argument of the integral in ΔE is

$$\frac{E_C}{V_C} - \frac{TS_C}{V_C} - \sum_{i=1}^{N_C} \frac{n_{Ci}}{V_C} \mu_{Li} - \left(\frac{v_{CL}}{v_C} \right) \left(\frac{E_L}{V_L} - \frac{TS_L}{V_L} \right)$$

$$+ \sum_{i=1}^{N_C} \frac{n_{Li}}{V_L} \left(\frac{v_{CL}}{v_C} \right) \mu_{Li} + \kappa_{CL} \left(E_I - TS_I - \sum_{i=2}^{N} \Gamma_i \mu_{Li} \right)$$

$$+ P_A - P_L \left(\frac{v_{CL}}{v_C} - 1 \right) \tag{37}$$

Using Eqs. 23, 24, and 28, Eq. 37 becomes

$$-P_C - \sum_{i=1}^{N_C} \frac{n_{Ci}}{V_C} \mu_{Ci} - \sum_{i=1}^{N_C} \frac{n_{Ci}}{V_C} \mu_{Li} + P_A + P_L + \gamma_{CL} \kappa_{CL} \tag{38}$$

At equilibrium, this quantity must be zero, so making use of Eq. 34 we obtain

$$\sum_{i=1}^{N_C} \frac{n_{Ci}}{V_C} (\mu_{Li} - \mu_{Ci}) = (\gamma_{CL} - f_{CL}) \kappa_{CL} \tag{39}$$

As Cahn[66] points out, the chemical potential change on moving between the solid and liquid phases, $\mu_{Li} - \mu_{Ci}$, must be the same for each component, or the composition of the crystal would have to change. Therefore, Eq. 39

reduces to

$$\mu_{Li} - \mu_{Ci} = v_C (\gamma_{CL} - f_{CL}) \kappa_{CL} \tag{40}$$

where v_C is the molar volume of the crystal. For a single-component system, this reduces to

$$\mu_L - \mu_C = v_C (\gamma_{CL} - f_{CL}) \kappa_{CL} \tag{41}$$

Eq. 41 indicates that the chemical potentials of the solid and liquid are not equal unless $f_{CL} = \gamma_{CL}$. This surprising result, which was first obtained by Gibbs,[65] has been discussed by Cahn[66] and Cammarata.[67]

Now we will apply this result to a crystal of ice growing from a brine. Taking water/ice to be component 1, the chemical potential of the liquid water is

$$\mu_{L1}(P_L, T) = \mu_{L1}(P_e, T_m) + \bar{V}_{L1}(P_L - P_e) - S_{L1}(T - T_m)$$
$$+ RT \ln(a_1) \tag{42}$$

where T_m is the melting point of the pure ice, P_e is the equilibrium vapor pressure, and a_1 is the activity of water in the solution. The chemical potential of the ice is

$$\mu_{C1}(P_C, T) = \mu_{C1}(P_e, T_m) + v_C (P_A + P_L + f_{CL}\kappa_{CL} - P_e)$$
$$- S_{C1}(T - T_m) \tag{43}$$

Using Eq. 40, and recognizing that $\mu_{L1}(P_e, T_m) = \mu_{C1}(P_e, T_m)$, we find

$$v_C (P_A + \gamma_{CL}\kappa_{CL}) = (\bar{V}_{L1} - v_C)(P_L - P_e) + (S_{L1} - S_{C1})(T - T_m)$$
$$+ RT \ln(a_1) \tag{44}$$

Note that if we assumed that $\gamma_{CL} = f_{CL}$ and wrote the pressure in the crystal as $P_C = P_A + P_L + \gamma_{CL}\kappa_{CL}$, as was done in Refs. 6 and 12 (among many others), we would obtain an expression identical to Eq. 44. However, that is not a valid procedure, and can lead to errors in more complicated cases.[66]

We will now examine some special cases. If the liquid is pure water, then $a_1 = 1$ and Eq. 44 becomes

$$v_C (P_A + \gamma_{CL}\kappa_{CL}) = (v_L - v_C)(P_L - P_e) + (S_L - S_C)(T_m - T) \tag{45}$$

where v_L is the molar volume of water.

For a macroscopic crystal ($\kappa_{CL} = 0$) of pure ice with no applied pressure ($P_A = 0$) in pure water, Eq. 44 reduces to the Clausius-Clapeyron equation:

$$P_L - P_e = \frac{S_{LI} - S_{CI}}{\overline{V}_{LI} - v_C}(T - T_m) \tag{46}$$

On the other hand, if the liquid/vapor menisci are flat ($P_L = P_e$), the water is pure, and there is no applied pressure on the crystal, Eq. 44 reduces to the Gibbs-Thomson equation:

$$\gamma_{CL}\kappa_{CL} = \frac{(S_{LI} - S_{CI})(T_m - T)}{v_C} \tag{47}$$

or, if the crystal is macroscopic, but a mechanical pressure is applied, then a similar shift in melting point occurs:

$$P_A = \frac{(S_{LI} - S_{CI})(T_m - T)}{v_C} \tag{48}$$

For a macroscopic crystal where $P_L = P_e$ and $P_A = 0$, Eq. 44 yields the shift in melting point by the solute[68]:

$$\ln(a_1) = -\frac{(S_{LI} - S_{CI})(T_m - T)}{RT} = \frac{\Delta H_m}{R}\left(\frac{1}{T_m} - \frac{1}{T}\right) \tag{49}$$

Suppose that a macroscopic ice crystal is not in contact with liquid water, but is exposed to the vapor of pure water in a capillary. The chemical potential of the ice is

$$\mu_C(P_C, T) = \mu_C(P_e, T_m) - S_C(T - T_m) \tag{50}$$

The chemical potential of the pure water in the capillary is found from Eq. 42:

$$\mu_L(P_L, T) = \mu_L(P_e, T_m) + v_L(P_L - P_e) - S_L(T - T_m) \tag{51}$$

The chemical potentials are equal at equilibrium and $\mu_C(P_e, T_m) = \mu_L(P_e, T_m)$, so in this case,

$$v_L(P_L - P_e) = (S_L - S_C)(T - T_m) \tag{52}$$

References

1. T. C. Powers, "The Air Requirement of Frost-Resistant Concrete," *Proc. Highway Res. Board*, **29**, 184–211 (1949).
2. G. G. Litvan, "Phase Transitions of Adsorbates: III. Heat Effects and Dimensional Changes in Nonequilibrium Temperature Cycles," *J. Colloid Interface Sci.*, **38** [1] 75–83 (1972).
3. J. J. Beaudoin and C. MacInnis, "The Mechanism of Frost Damage in Hardened Cement Paste," *Cem. Concr. Res.*, **4** [2] 139–147 (1974).
4. T. C. Powers and R. A. Helmuth, "Theory of Volume Changes in Hardened Portland-Cement Paste during Freezing," *Proc. Highway Res. Board*, **32**, 285–297 (1953).
5. J. Israelachvili, *Intermolecular and Surface Forces*, 2nd ed. Academic, London, 1992.
6. G. W. Scherer, "Crystallization in Pores," *Cem. Concr. Res.*, **29** [8] 1347–1358 (1999); G. W. Scherer, "Reply to Discussion of Crystallization in Pores," *Cem. Concr. Res.*, **30** [4] 673–675 (2000).
7. A. E. Corte, "Vertical Migration of Particles in Front of a Moving Freezing Plane," *J. Geophys. Res.*, **67** [3] 1085–1090 (1962).
8. K. A. Jackson, D. R. Uhlmann, and B. Chalmers, "Frost Heave in Soils," *J. Appl. Phys.*, **37** [2] 848–852 (1966).
9. R. A. Helmuth, "Discussion of the Paper 'Frost Action in Concrete' by P. Nerenst"; pp. 829–833 in *Proceedings of the 4th International Congress on Chemistry of Cement*, vol. 2. NBS Monograph 43. National Bureau of Standards, Washington, D.C., 1962.
10. G. Fagerlund, "Studies of the Destruction Mechanism at Freezing of Porous Materials"; in *Contributions to Fondation Française d'Études Nordiques, VIe Congres Int.: Les problèmes posés par la gélifraction. Recherches fondamentales et apppliquées* (French Foundation for Nordic Studies, 6th International Congress on Problems Raised by Freezing. Fundamental and Applied Research). (Le Havre, 23–25 April 1975.)
11. J. Thomson, "Theoretical Considerations on the Effect of Pressure in Lowering the Freezing Point of Water," *Trans. Roy. Soc. Edinburgh*, **16** [5] 575–580 (1848–1849).
12. D. H. Everett, "The Thermodynamics of Frost Damage to Porous Solids," *Trans. Faraday Soc.*, **57**, 1541–1551 (1961).
13. M. Buil and J. Aguirre-Puente, "Thermodynamic and Experimental Study of the Crystal Growth of Ice"; pp. 1–7 in *Proceedings of the ASME Winter Annual Meeting* (1981) [81-WA/HT-69].
14. G. W. Scherer, "Freezing Gels," *J. Non-Cryst. Solids*, **155**, 1–25 (1993).
15. R. Defay and I. Prigogine, *Surface Tension and Adsorption*. Wiley, New York, 1966.
16. M. Brun, A. Lallemand, J. F. Quinson, and C. Eyraud, "A New Method for the Simultaneous Determination of the Size and the Shape of Pores: The Thermoporometry," *Thermochimica Acta*, **21**, 59–88 (1977).
17. W. B. Hillig and D. Turnbull, "Theory of Crystal Growth in Undercooled Pure Liquids," *J. Chem. Phys.*, **24**, 914 (1956).
18. R. A. Helmuth, "Capillary Size Restrictions on Ice Formation in Hardened Portland Cement Pastes,"; pp. 855–869 in *Proceedings of the 4th International Congress on Chemistry of Cement*, vol. 2. NBS Monog. 43. National Bureau of Standards, Washington, D.C., 1962.
19. H. Liu, L. Zhang, and N. A. Seaton, "Analysis of Sorption Hysteresis in Mesoporous

Solids Using a Pore Network Model," *J. Colloid Interface Sci.*, **156**, 285–293 (1993); erratum, **162**, 265 (1994).

20. A. J. Katz and A. H. Thompson, "Prediction of Rock Electrical Conductivity from Mercury Injection Measurements," *J. Geophys. Res.*, **92** [B1] 599–607 (1987).

21. A. J. Katz and A. H. Thompson, "Quantitative Prediction of Permeability in Porous Rock," *Phys. Rev. B*, **34** [11] 8179–8181 (1986).

22. P. V. Hobbs, *Ice Physics*. Clarendon, Oxford, 1974.

23. M. A. Biot, "Theory of Propagation of Elastic Waves in a Fluid-Saturated Porous Solid. I. Low-Frequency Range," *J. Acoustical Soc. Am.*, **28** [2] 168–178 (1956).

24. G. W. Scherer, "Measuring Permeability of Rigid Materials by a Beam-Bending Method: I. Theory," *J. Am. Ceram. Soc.*, **83** [9] 2231–2239 (2000); erratum, *J. Am. Ceram. Soc.*, **87** [8] 1612–1613 (2004).

25. W. Vichit-Vadakan and G. W. Scherer, "Measuring Permeability of Rigid Materials by a Beam-Bending Method: II. Porous Vycor," *J. Am. Ceram. Soc.*, **83** [9] 2240–2245 (2000); erratum, *J. Am. Ceram. Soc.*, **87** [8] 1614 (2004).

26. S. Chatterji, "Aspects of the Freezing Process in a Porous Material–Water System. Part 1. Freezing and the Properties of Water and Ice," *Cem. Concr. Res.*, **29**, 627–630 (1999).

27. J. P. Ciardullo, D. J. Sweeney, and G. W. Scherer, "Thermal Expansion Kinetics: Method to Measure Permeability of Cementitious Materials: IV, Effect of Thermal Gradients," accepted for publication in *J. Am. Ceram. Soc.*

28. W. Vichit-Vadakan and G. W. Scherer, "Measuring Permeability of Rigid Materials by a Beam-Bending Method: III. Cement Paste," *J. Am. Ceram. Soc.*, **85** [6] 1537–1544 (2002); erratum, *J. Am. Ceram. Soc.*, **87** [8] 1615 (2004).

29. E. J. Sellevold and D. H. Bager, "Some Implications of Calorimetric Ice Formation Results for Frost Resistance Testing of Cement Products"; pp. 1–27 in *International Colloquium on Frost Resistance of Concrete*. Wien, 1980.

30. C. le Sage de Fontenay and E. J. Sellevold, "Ice Formation in Hardened Cement Paste, I. Mature Water-Saturated Pastes"; pp. 425–438 in *Durability of Building Materials and Components*. Edited by P. J. Sereda and G. G. Litvan. ASTM STP 691. American Society for Testing and Materials, Philadelphia, 1980.

31. Z. P. Bazant, J.-C. Chern, A. M. Rosenberg, and J. M. Gaidis, "Mathematical Model for Freeze-Thaw Durability of Concrete," *J. Am. Ceram. Soc.*, **71** [9] 776–783 (1988).

32. G. G. Litvan, "Phase Transitions of Adsorbates: VI, Effect of Deicing Agents on the Freezing of Cement Paste," *J. Am. Ceram. Soc.*, **58** [1–2] 26–30 (1958).

33. G. W. Scherer, J. Chen, and J. Valenza, "Method of Protecting Concrete from Freeze Damage," U.S. Patent 6 485 560 (Nov. 26, 2002).

34. S. Chatterji, "Freezing of Air-Entrained Cement-Based Materials and Specific Actions of Air-Entraining Agents," *Cem. Concr. Res.*, **25**, 759–765 (2003).

35. R. C. Mielenz, V. E. Volkodoff, J. E. Backstrom, and H. L. Flack, "Origin, Evolution, and Effects of the Air Void System in Concrete — Part 1: Entrained Air in Unhardened Concrete," *ACI Mater. J.*, **55**, 95–121 (1958).

36. D. J. Corr, P. J. M. Monteiro, and J. Bastacky, "Microscopic Characterization of Ice Morphology in Entrained Air Voids," *ACI Mater. J.*, **99-M18** [March-April] 190–195 (2002).

37. G. Fagerlund, "The International Cooperative Test of the Critical Degree of Saturation

Method of Assessing the Freeze/Thaw Resistance of Concrete," *Materiaux Constructions*, **10** [58] 230–251 (1977).

38. G. Fagerlund, "Predicting the Service Life of Concrete Exposed to Frost Action through a Modelling of the Water Absorption Process in the Air-Pore System"; pp. 503–537 in *The Modelling of Microstructure and Its Potential for Studying Transport Properties and Durability*. Edited by H. Jennings. Kluwer, Amsterdam, 1996.

39. J. Happel and H. Brenner, *Low Reynolds Number Hydrodynamics*. Martinus Nijhoff, Dordrecht, 1986.

40. J. Marchand, R. Pleau, and R. Gagné, "Deterioration of Concrete Due to Freezing and Thawing"; pp. 283–354 in *Materials Science of Concrete IV*. Edited by J. Skalny and S. Mindess. American Ceramic Society, Westerville, Ohio, 1995.

41. M. Pigeon and R. Pleau, *Durability of Concrete in Cold Climates*. E&FN Spon, London, 1995.

42. S. Lindmark, "Mechanisms of Salt Frost Scaling of Portland Cement-Bound Materials: Studies and Hypothesis," Ph.D. thesis (Report TVBN 1017), Lund Institute of Technology, Lund, Sweden, 1998.

43. J. Marchand, M. Pigeon, D. Bager, and C. Talbot, "Influence of Chloride Solution Concentration of Salt Scaling Deterioration of Concrete," *ACI Mater. J.*, (July–August 1999), pp. 429–435.

44. G. J. Verbeck and P. Klieger, "Studies of 'Salt' Scaling of Concrete," *Highway Research Board Bull.*, **150**, 1–17 (1957).

45. S. Jacobsen, "Scaling and Cracking in Unsealed Freeze/Thaw Testing of Portland Cement and Silica Fume Concretes," thesis report 1995:101, Norwegian Institute of Technology, Trondheim, 1995.

46. E. J. Sellevold and T. Farstad, "Frost/Salt Testing of Concrete: Effect of Test Parameters and Concrete Moisture History," *Nordic Concr. Res.*, no. 10, 121–138 (1991).

47. W. Studer, "Internal Comparative Tests on Frost-Deicing-Salt Resistance"; pp. 175–187 in *International Workshop on Resistance of Concrete to Scaling Due to Freezing in the Presence of Deicing Salts*. Centre Recherche Interuniv. Beton, University of Sherbrooke, University Laval, Québec (August 1993).

48. A. Rösli and A. B. Harnik, "Improving the Durability of Concrete to Freezing and Deicing Salts"; pp. 464–473 in *Durability of Building Materials and Components*. Edited by P. J. Sereda and G. G. Litvan. ASTM STP-691. American Society for Testing and Materials, Philadelphia, 1980.

49. J. Stark and H. M. Ludwig, "Freeze-Thaw and Freeze-Deicing Salt Resistance of Concretes Containing Cement Rich in Granulated Blast-Furnace Slag"; paper 4-iv-035 in *Proceedings of the 10th International Congress on the Chemistry of Cement*, vol. 4. SINTEF, Trondheim, Norway, 1997.

50. M. A. Biot, "General Theory of Three-Dimensional Consolidation," *J. Appl. Phys.*, **12**, 155–164 (1941).

51. G. W. Scherer, "Thermal Expansion Kinetics: Method to Measure Permeability of Cementitious Materials: I, Theory," *J. Am. Ceram. Soc.*, **83** [11] 2753–2761 (2000); erratum, *J. Am. Ceram. Soc.*, **87** [8] 1609–1610 (2004).

52. J. J. Valenza and G. W. Scherer, "Mechanism of Salt Scaling," in preparation.

53. J. J. Valenza II and G. W. Scherer, "Measuring Permeability of Rigid Materials by a

Beam-Bending Method: V. Cement Paste Plates," *J. Am. Ceram. Soc.*, **87** [10] 1927–1931 (2004).

54. J. Stark and S. Stürmer, "Investigation of Compatibility of Cements with Several Salts"; paper 4-iv-031 in *Proceedings of the 10th International Congress on the Chemistry of Cement*, vol. 4. SINTEF, Trondheim, Norway, 1997.

55. S. Y. Hong and F. P. Glasser, "Alkali Binding in Cement Pastes. Part I. The C-S-H Phase," *Cem. Concr. Res.*, **29**, 1893–1903 (1999).

56. H. Viallis, P. Faucon, J. C. Petit, and A. Nonat, "Interaction between Salts (NaCl, CsCl) and Calcium Silicate Hydrates (C-S-H)," *J. Phys. Chem. B*, **103**, 5212–5219 (1999).

57. U. Wiens and P. Schiessl, "Chloride Binding of Cement Paste Containing Fly Ash"; paper 4-iv-016 in *Proceedings of the 10th International Congress on the Chemistry of Cement*, vol. 4. SINTEF, Trondheim, Norway, 1997.

58. A. G. Kalinichev and R. J. Kirkpatrick, "Molecular Dynamics Modeling of Chloride Binding to the Surfaces of Calcium Hydroxide, Hydrated Calcium Aluminate, and Calcium Silicate Phases," *Chem. Mater.*, **14**, 3539–3549 (2002).

59. S. Timoshenko, "Analysis of Bimetal Thermostats," *J. Opt. Soc. Am.*, **11**, 233–255 (1925).

60. S. T. Gulati and H. Hagy, "Analysis and Measurement of Glue-Spall Stresses in Glass-Epoxy Bonds," *J. Am. Ceram. Soc.*, **65** [1] 1–6 (1982); correction, **65** [6] 320 (1982).

61. T. A. Hammer and E. J. Sellevold, "Frost Resistance of High-Strength Concrete"; pp. 457–487 in *2nd Symposium on High Strength Concrete*. SP-121. American Concrete Institute, Detroit, 1990.

62. J. J. Valenza II and G. W. Scherer, "Mechanism for Salt Scaling of a Cementitious Surface," accepted for publication in *Concr. Sci. Eng.*

63. J. J. Valenza II and G. W. Scherer, "Salt Scaling — A Comprehensive Review of the Phenomenology and Proposed Mechanisms," submitted to *Cem. Concr. Res.*

64. T. Ye, Z. Suo, and A. G. Evans, "Thin Film Cracking and the Roles of Substrate and Interface," *Int. J. Solid Structures*, **29**, 2639–2648 (1992).

65. J. Willard Gibbs, *The Scientific Papers: I. Thermodynamics*. Dover, New York. Pp. 316–317.

66. J. W. Cahn, "Surface Stress and the Chemical Equilibrium of Small Crystals — I. The Case of the Isotropic Surface," *Acta Metall.*, **28**, 1333–1338 (1980).

67. R. C. Cammarata, "Surface and Interface Stress Effects in Thin Films," *Prog. Surf. Sci.*, **46**, 1–37 (1994).

68. F. T. Wall, *Chemical Thermodynamics*. Freeman, San Francisco, 1965. Pp. 350–353.

Early Age Flexural Strength: The Role of Aggregates and Their Influence on Maturity Predictions

A. Barde, G. Mazzotta, and J. Weiss

This article discusses early age flexural strength development in concrete, specifically investigating the role of aggregates and their impact in maturity predictions. The maturity method is becoming more widely used by the construction industry to signal when certain construction operations (e.g., opening to traffic or removing formwork) can be performed. In addition, the maturity method is used in various computer programs to simulate how physical properties develop during hydration. The maturity method is based on the concept that strength (or mechanical property) development is proportional to the extent of chemical reaction (i.e., hydration) that has taken place. It is commonly assumed that the extent of chemical reaction (i.e., the degree of hydration) is a unique function of the product of time and temperature. It is the hypothesis of this work that aggregates can alter this relationship specifically influencing the early age relationship between maturity and flexural (or tensile) strength. To verify this hypothesis the nonevaporable water (i.e., the degree of hydration) was related to the flexural strength development in paste, mortar, and concrete specimens. A linear relationship was observed between the flexural strength and the nonevaporable water for paste specimens, while a bilinear response was observed for both mortar and concrete. The knee point of this bilinear behavior corresponds to the time at which the majority of the aggregates begin to fracture. At very early ages (i.e., less than 2.5 days in this study) the flexural failure behavior is dominated by the paste or bond failure, while at later ages flexural failure is dominated by aggregate failure.

Introduction

The rate at which the mechanical properties of concrete develop is strongly influenced by the temperature of the concrete and its surrounding environment. The maturity method has been proposed as an approach to estimate mechanical property development in concrete with variable temperature histories.[1] The basic concept behind the maturity method is that the product of the time and temperature can be used as a unique index to describe the rate of chemical reaction (i.e., degree of hydration) that has taken place. This maturity index can be related to mechanical property development (i.e., strength or stiffness) and used to describe strength and stiffness under the specific temperature history to which the structure is exposed.

The foundation of the maturity method is rooted in the assumption that property development is proportional to the extent of the hydration reaction.[2] An excellent review of the historical development and application of various maturity approaches has been provided by Carino.[3] In one approach the strength of the concrete can be described using a hyperbolic master curve

$$S = S_\infty \left[\frac{k_t (t - t_0)}{1 + k_t (t - t_0)} \right] \qquad (1)$$

where S_∞ is the long-term strength, k_t is the rate constant, and t_0 is the offset time, which corresponds to the beginning of strength development.[1] The rate constant accounts for the effects of temperature on property development and is frequently described using the Arrhenius equation.[1,4] Inherently this approach is based on the assumption that temperature changes alter only the rate of the hydration reaction.

Despite widespread use, the maturity method must be used judiciously. The precise prediction of strength development using the maturity method requires that the mixture design and constituent materials used to develop the relationship between the maturity and strength are similar to the concrete for which the properties are to be predicted. Graveen et al.[5] recently illustrated that even the slight variations in mixture proportions that can occur during construction should be considered for accurate maturity predictions.

It was hypothesized[5] that aggregates play a key role in flexural strength development, and after a specific degree of hydration is achieved the presence of aggregates can alter the maturity–strength gain relationship. Guinea et al.[6] showed that the flexural strength of concrete can significantly vary as a function of quality of paste/aggregate interface. To capture the influence of the properties of each phase, researchers have proposed models to explain aggregate stiffness, strength, and interface energy on the properties of concrete.[7] Wu et al.[8] demonstrated that as the strength of the concrete increases, the aggregates become increasingly important. This chapter builds on these observations and describes an investigation that was conducted to illustrate how the rate constant may be influenced by more than just the hydration reaction of the paste phase depending on the specific property that is being determined.

This article describes flexural strength development in paste, mortar, and concrete. The specific intention of this paper is to demonstrate that the use of the maturity method for the prediction of material properties that are dominated by mode I fracture may be complicated by the aggregate phase when it begins to fracture. It is shown that the flexural strength gain is described by the hydration of the paste phase at early ages; however, it is primarily governed by aggregates after a certain degree of hydration is reached. The relationship between flexural strength and degree of hydration exhibits a bilinear behavior when aggregates are present. It is believed that by understanding the composite nature of early age concrete, strength gain and fracture toughness can be better predicted for use in early age damage and failure modeling.

Experimental Program

This study investigates the flexural strength versus maturity relationship for various cementitious composites. Three series of tests were conducted as described in Table I. In the first series the water/cement (w/c) ratio was maintained constant while the aggregates were varied to include a paste (0% aggregate), a mortar (55% fine aggregates by volume), and a concrete (34% of fine aggregates and 36% coarse aggregates). In the second series the w/c was varied for paste and mortar specimens with w/c = 0.30, 0.36, 0.42, and 0.50. In the third series, the type of aggregate was varied to include single size granite and limestone aggregates with w/c = 0.30.

A series of tests was performed to better understand how aggregates influence early age mechanical properties. Flexural strength tests were performed to obtain a measure of mechanical property development while ultrasonic pulse velocity measurements were taken to quantify the rate of property development without the influence of aggregate fracture. Loss on ignition (LOI) tests were conducted to provide a measure of the extent of chemical reaction (nonevaporable water or degree of hydration) at each maturity measurement.

The following sections describe the mixture proportions in greater detail. In addition, information is provided about the materials used for this study. Experimental procedures are described along with details of the specimen geometry and specimen preparation.

The mixture proportions used in this research are shown in Table I. Type I ordinary portland cement was used for all mixtures (60% C_3S, 13.5% C_2S, 8.2% C_3A; Na_2O equivalent of 0.54% and fineness of 360 m^2/kg). In the first series, the aggregate content was varied to make paste, mortar, and concrete specimens. The mortar mixture was designed so that it was similar

Table I. Mixture proportions

Test series	Mixture	w/c	Cement content (kg/m³)	Vol% of aggregate FA	CA	Total	Experiment performed
I. Comparison of paste, mortar, and concrete	Paste	0.42	1355				Flexural strength and ultrasonic pulse velocity
	Mortar	0.42	561	55		55	
	Concrete	0.42	320	34	36	70	
II. Influence of water/cement ratio	Paste	0.30	1620				Flexural strength and loss on ignition
		0.36	1475				
		0.42	1355				
		0.50	1223				
	Mortar	0.30	728	55		55	
		0.36	664	55		55	
		0.42	561	55		55	
		0.50	550	55		55	
III. Influence of aggregate types	Paste	0.30	1620				Flexural strength
	Granite mortar	0.30	1133		30	30	
	Limestone mortar	0.30	1133		30	30	

Note: All mixtures contained 2.44 mL of a water-reducing admixture per kilogram of cement and 145 mL of an air-entraining admixture per cubic meter of concrete.

to the concrete mixture with the coarse aggregates removed. The paste specimens were similar to mortar and concrete specimens with the aggregates removed. In second series, the w/c ratio was varied to include paste and mortar specimens with w/c = 0.30, 0.36, 0.42, and 0.50. In the third series granite and limestone aggregates were used with w/c = 0.30. The aggregates that passed through the standard 9.5 mm sieve and were retained on standard 4.75 mm sieve were used. It should be noted that the aggregates were selected so that the strengths were considerably different to identify the influence of aggregates on flexural strength–maturity relationship development. These mixtures were designed with only paste and single size aggregates to minimize the influence of sand and aggregate size distribution that could further complicate the interpretation of the results.

Standard practices for making concrete test specimens (ASTM C-192) were followed. The concrete was mixed in a pan mixer, placed in the forms, rodded, vibrated using a plate vibrator, and finished with a steel trowel. The molds were then covered with wet burlap. The specimens tested before an age of 24 h were demolded 10 min before the tests, while the specimens tested after an age of 24 h were demolded and cured in a temperature-controlled moist curing room (23 ± 1°C, 98% RH) until the time of test. The temperature of all mixtures was monitored to compute the time-temperature history for each mixture.

Nonevaporable Water

Nonevaporable water content (w_{nevap}) was determined in accordance with ASTM C-114. Paste specimens (w/c = 0.30, 0.36, 0.42, 0.50) were used for the determination of w_{nevap}. At each age, approximately 2 g of paste was ground in a mortar and pestle and immediately placed in an acetone solution to arrest the hydration reaction. The specimens remained in acetone for 24 h before being dried in air for 2 h to allow excess acetone to evaporate. The specimens were then placed in an oven at 105°C for 24 h. The specimens were then ignited to 1050°C for 12 h and the differences between the weights (105–1050°C) were obtained to calculate w_{nevap} using Eq. 2.[9]

$$w_{nevap} = \frac{W_{105°} - W_{1050°}}{W_{105°}} \times 100 \tag{2}$$

where $W_{105°}$ is the weight of sample at 105°C and $W_{1050°}$ is the weight of the sample after heating to 1050°C.

Flexural Strength

Two different specimen geometries were used for flexural strength measurement. Smaller specimen geometries were used for the paste and mortar specimens with a length of 300 mm (12 in.) and a cross section of 76 × 76 mm (3 × 3 in.). Larger specimen geometries were prepared for the concrete mixture with a length of 530 mm (21 in.) and a cross section of 152 × 152 mm (6 × 6 in.). Flexural specimens were tested in third point loading as specified in ASTM C-78 with a span-to-depth ratio of 3. For the smaller specimens a loading rate of 150 psi/min was used while the larger specimens tested in displacement control used a loading rate of 0.0005 mm/min. The load deflection data was used to compute the elastic modulus *(E)* of the beams. (Deflection was computed using the average center point displacement of the beam as measured using LVDTs attached to a yoke on either side of the beam.) In addition, ultrasonic pulse velocity (UPV) tests were performed using two 100 mm (4 in.) diameter and 200 mm (8 in.) height cylinders. The elasticity modulus obtained from UPV and load deflection tests were cured under similar conditions to the curing conditions in which the concrete beams were stored.

Maturity Measurement

The time-temperature history of each mixture was recorded using two thermocouples for paste, mortar, and concrete mixture using a Campbell Scientific CR10X measurement and control system. The time-temperature histories were then used to calculate the maturities for the corresponding mixtures using Eq. 3.[1]

$$M = \int_{t_0}^{t} \left[k(T) dt \right] \tag{3}$$

where M is the maturity, t is the time, t_0 is the offset time, and $k(T)$ is the rate constant at temperature T.

Experimental Results

Series I: Comparison of Paste, Mortar, and Concrete

The flexural strengths of the paste, mortar, and concrete specimens were obtained in the first test series at various ages, and the strength prediction

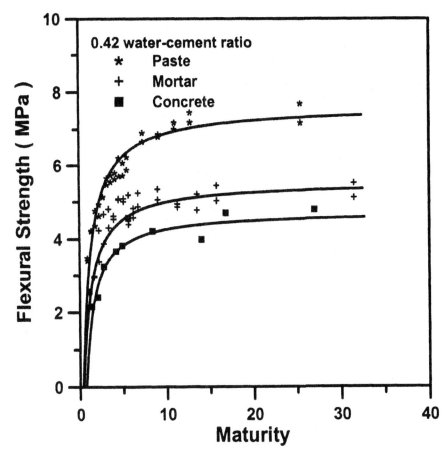

Figure 1. Flexural strength maturity master curve for Series I: Paste, mortar, and concrete with w/c = 0.42.

curves (Eq. 1) were prepared using the maturity method as outlined by Carino.[1] Figure 1 shows the variation of flexural strength development for paste, mortar, and concrete mixtures as a function of maturity (w/c = 0.42).

Though the early age flexural strength looks similar for paste, mortar, and concrete specimens, it should be noted that it was different at later ages. It is the hypothesis of this chapter that these differences may be attributed to the presence of aggregates. It should be noted that the hyperbolic strength curve is able to predict the strength development for all these series (paste, mortar, and concrete). However it was observed that the rate constants for the paste, mortar, and concrete mixtures were not the same.

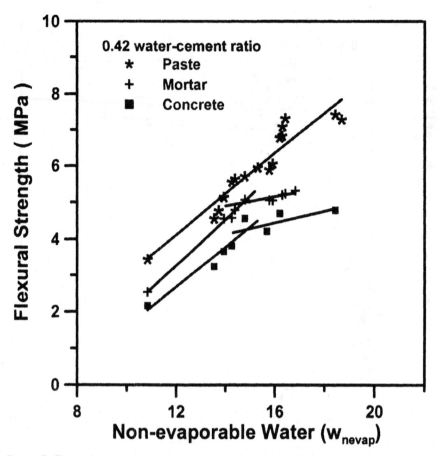

Figure 2. Flexural strength versus nonevaporable water (proportional to degree of hydration) for Series I: Paste, mortar, and concrete with w/c = 0.42.

Figure 2 shows the flexural strength development for concrete, mortar, and paste as a function of nonevaporable water (w_{nevap}, which is proportional to the degree of hydration). The relationship between nonevaporable water was bilinear for strength development of mortar and concrete while the paste specimen still showed a linear response. The slope between the strength and nonevaporable water is similar at low nonevaporable water contents (i.e., low degree of hydration) for paste, mortar, and concrete, although it should be noted that the flexural strength of the mortar specimens was less than that of the paste for the same nonevaporable water content, and the concrete specimen was even lower. It is currently believed that

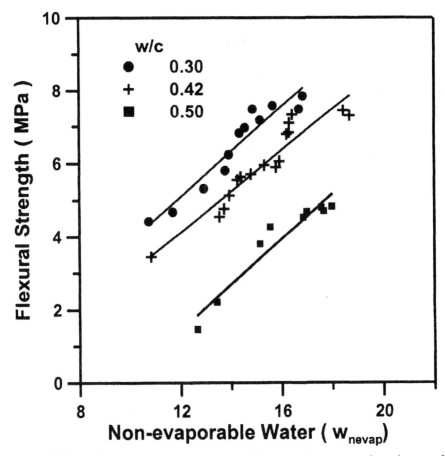

Figure 3. Flexural strength versus nonevaporable water (proportional to degree of hydration) for Series II: Paste with w/c = 0.30, 0.42, and 0.50.

this may be due to either the presence of very weak aggregates or an interface that is weaker than the strength of the paste. Further research is needed to clarify the exact reason for the lower strength in the mortar and concrete systems.

Series II: Influence of Water/Cement Ratio

The nonevaporable water content was determined for paste specimens with various water/cement ratios and correlated with flexural strength gain. Figure 3 illustrates the strength versus nonevaporable water in the paste specimens with varying water/cement ratios. It can be observed that the strength

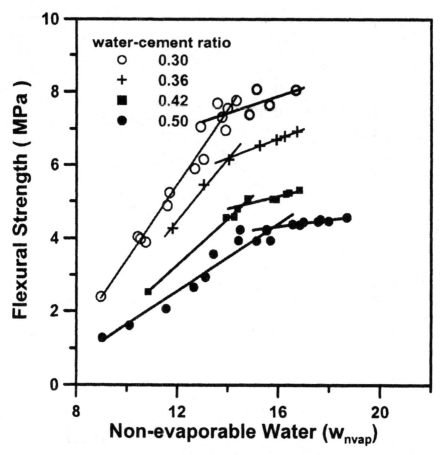

Figure 4. Flexural strength versus nonevaporable water (proportional to degree of hydration) for Series II: Mortar with w/c = 0.30, 0.36, 0.42, and 0.50).

development is linearly proportional to the nonevaporable water, indicating that the strength development in the paste is directly related to the extent of the hydration reaction that has taken place. As such it appears that the application of the maturity method to describe the rate of strength development of paste would be relatively straightforward with a change in the time-temperature history would result in a change in the degree of hydration (i.e., nonevaporable water content).

The flexural strength development of the mortar specimens with a varying w/c value as a function of nonevaporable water is shown in Fig. 4. Unlike the paste behavior (Fig. 3), the flexural strength of the mortar specimens (Fig. 4) again showed a bilinear relationship with nonevaporable water (DOH). The knee point in the bilinear mortar curves (point of change in slope) is related to the behavior of aggregates, representing the critical DOH where a single volume of aggregate begin to fracture. It is worth noting that the initial rate of flexural strength development in the mortar is very similar to the rate of flexural strength development of paste until the knee point is achieved. It is believed that at early ages the crack propagates around the aggregates. Therefore, at low DOH the failure is predominantly dependent on the paste strength governed by the hydration reaction. The presence of aggregates does not appear to substantially affect the rate of strength development at these early ages and only slightly reduces the strength. This reduction in strength is presumably due to either the fracture of a small portion of aggregates or the potential low strength (high porosity) of the interfacial region around the aggregates at early ages.

Figures 2 and 4 indicate that after the knee point, the rate of flexural strength development is significantly lower than the initial rate of strength development. (Recall that the initial rate of strength development in the mortar and concrete specimens is approximately equal to the rate of flexural strength development in the paste specimens.) It should be noted that the knee point appears to correspond with the time at which the visual observations of the crack surfaces show fractured aggregates. Further evidence is presented later in this paper for a more quantitative assessment of the fraction of the aggregate that has fractured at each stage of the test.

The time when the aggregates began to fracture, as determined using visual observations, for all these mixtures appeared to correspond to the knee point. It should be noted that the aggregates used in all of the mixtures (i.e., at all w/c ratios) were similar (i.e., the same source, gradation, and volume). The knee point that was observed in lower strength (higher w/c ratio) materials appears to illustrate that a greater degree of hydration is required to cause the aggregate to fracture. As expected, the mixture with a lower w/c ratio (i.e., 0.30) appears to develop strength or interface bonds more rapidly, therefore the lower w/c ratio mixture reaches the knee point at an earlier age than mixtures with a higher w/c ratio. It should be noted that the rate of the flexural strength development for each paste mixture was similar; however, the rate of strength development of mortar specimens reduced from mixture with 0.30 w/c ratio to a mixture with a 0.50 w/c ratio.

The fractured surfaces of the concrete beams were visually examined to determine the proportion of aggregates fractured at each testing age. Figure

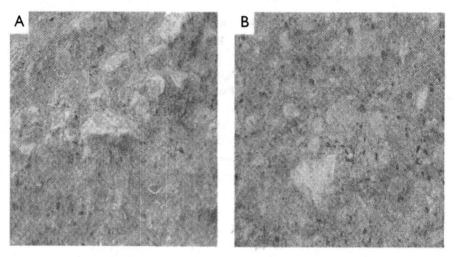

Figure 5. Photos of fractured surface; AFD measurements at (a) early age (0.5 days before knee point) and (b) later age (3 days after knee point).

5 shows an example of two fracture surfaces from beams tested at an age of 12 and 72 h. It can be seen in Fig. 5(a) (the specimen tested at 12 h) that the crack propagates around the coarse aggregate, while Fig. 5(b) (the specimen tested at 72 h) shows the crack propagating through the coarse aggregate.

To better quantify the visual observations the aggregate fracture density (AFD) was calculated for each fractured surface. The AFD was computed as the ratio of the number of fractured aggregates to the total number of coarse aggregates in the fractured surface. All coarse aggregates (approximately greater than 4.75 mm) were counted and classified as either unfractured or fractured.

The aggregate fracture density was observed, supporting the hypothesis that the knee point behavior corresponds with aggregate fracture. A low number of aggregates were fractured at the early age where the failure was dominated by the paste matrix or the bond. At an early age (i.e., 12 h) the AFD was only 16%, which corresponds with the fact that the majority of the aggregates were undamaged as the crack propagated around them. After the knee point was reached a more substantial percentage of the aggregates were observed to have fractured. For example, at an age of 2 days, the AFD was observed to be 64%, which showed that most of the aggregates were fractured. And at 6 and 28 days the AFD was calculated as 82% and 92%, respectively, verifing that almost all of the aggregates were fractured.

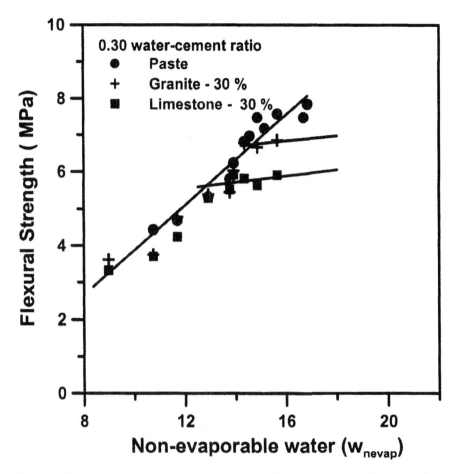

Figure 6. Flexural strength versus the nonevaporable water content (for paste, limestone, and granite).

Series III: Variation in Aggregate Types

To better understand how the knee point was related to the effects of strength of aggregates, two additional mixtures were prepared in the third series. The mixtures were prepared using a constant w/c ratio and a single size granite or limestone aggregate using the mixture proportions provided in Table I. Again flexural beams were prepared (with a length of 300 mm [12 in.] and cross section of 76×76 mm [3×3 in.]) and tested in flexure.

Figure 6 shows the flexural strength behavior of these beams as a function of nonevaporable water along with the flexural strength behavior of

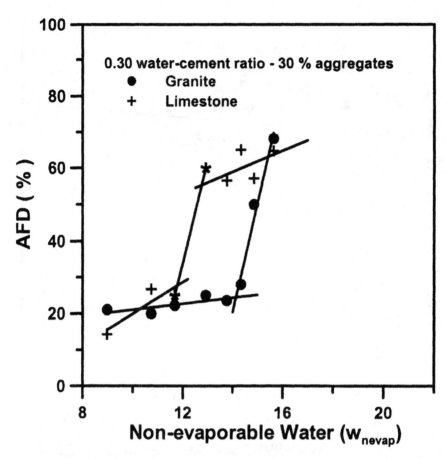

Figure 7. Aggregate fracture density of a concrete containing either granite or lime-stone aggregates.

paste specimens. It can be seen that the knee point occurs at different nonevaporable water contents for the mixtures with the two different aggregates. It should be noted that the initial strength gain was similar for the mixtures with aggregates and the pure paste mixture up to the knee point. After the knee point, however, the aggregates influence the strength at failure. Again the AFD was calculated for these mixtures as shown in Fig. 6. Since limestone is weaker than granite, it was expected that the knee point for the limestone mixture would occur at a lower paste maturity (i.e., lower degree of hydration), which can be observed in Figure 7.

Elastic Modulus Behavior

The elastic modulus was calculated for the concrete beams using the load displacement data (measured using the average of two LVDTs mounted on either side of a yoke on the beam). The slope of the load displacement response was used to calculate the elastic modulus using the standard displacement equation for simply supported elastic bend under third point loading. In addition, the elastic modulus of the same mixture was calculated using information from the ultrasonic pulse velocity.

Figure 8 shows the variation in elastic modulus (of the beams) as a function of the nonevaporable water (i.e., degree of hydration). The elastic modulus varies as a linear function of the nonevaporable water and no knee point is observed. Even though this initially may not seem logical in view of the previous discussions on the knee point observed for the same beams in describing the flexural strength, it should be noted that the elastic modulus depends on the behavior of the specimen under very low levels of loading. At these low load levels it can be assumed that the concrete has not developed significant damage and therefore that the aggregates have not fractured, and as such do not influence the elastic modulus development. This approach may explain the observations of Rostasy et al.,[10] who note that the elastic modulus increases at a much different rate than flexural strength. However, further work is being conducted to ascertain whether this is the source of these differences.

A Preliminary Correction for Maturity Predictions That Consider Aggregates

A very preliminary approach is presented below to incorporate the behavior of aggregates in the prediction of a maturity-strength relationship for concrete. In this discussion an approach can be used that considers the concrete as a composite material of a paste matrix phase and an aggregate phase. It could be imagined that the flexural strength of a two-phase composite could be approximated using a standard composite law of mixtures approach as shown in Eq. 4.

$$\sigma_{r\text{-conc}} = \sigma_{paste} V_{paste} + \sigma_{agg} V_{agg} \tag{4}$$

where $\sigma_{r\text{-conc}}$ is the flexural strength of concrete, σ_{paste} is the strength of the paste, σ_{agg} is the strength of aggregates, V_{paste} is the volume fraction of the paste, and V_{agg} is the volume fraction of the aggregates. (Note that in this

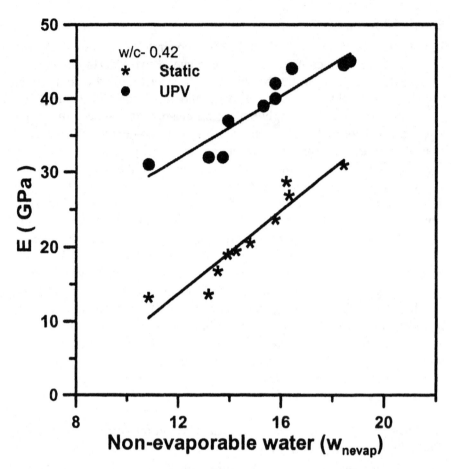

Figure 8. Elastic modulus versus nonevaporable water (proportional to degree of hydration) for Series I concrete with w/c = 0.42.

approach it is assumed that the elastic stiffnesses are similar enough that they do not require substantial corrections like those more commonly proposed in fiber-reinforced composites.[11] However, based on the previous discussion this approach would need to be modified to reflect the fact that the aggregate contribution changes as a function of the degree of hydration.

It can be assumed that the flexural strength of concrete at early ages has a failure that is dominated by the paste phase (i.e., before the knee point). The strength development of the paste phase can be predicted using the hyperbolic strength development curve as previously described. At early ages the aggregates are stronger than the paste phase, therefore the fracture takes place around the aggregate entirely through the paste phase. As such the fracture surface shows that the crack is essentially 100% through the paste phase. If we neglect the fact that the fracture path will be longer than the path straight across a paste specimen, the flexural strength at early ages can therefore be approximately predicted using Eq. 5.

$$\sigma_{r\text{-conc}} \quad \sigma_{r\text{ paste}} \tag{5}$$

where $\sigma_{r\text{-paste}}$ is the flexural strength predicted using maturity method of the same w/c ratio paste system.

At later ages (i.e., after the knee point) the failure is attributed to the contribution of both paste matrix and aggregates to the strength of concrete (Eq. 4) and can be estimated using the approach described in Eq. 4, which has been rearranged in Eq. 6.

$$\sigma_{r\text{-conc}} = \sigma_{r\text{-paste}} - \left(\sigma_{\text{paste}} - \sigma_{\text{agg}}\right) V_{\text{agg}} \tag{6}$$

Assuming a tensile strength of 3.80 MPa for a limestone aggregate, the results obtained from the experiments and from Eq. 6 are similar and appear to capture the later age behavior (after the knee point).

Though it has been proposed that early age strength is a function of paste behavior alone, it should be noted that at early ages the mixtures with aggregate inclusions show slightly lower values of flexural strength as compared to the corresponding flexural strength values of paste specimens at similar ages. This behavior is currently thought to be attributed to stress concentrations around the aggregates and an increasing volume of the interfacial transition zone material that may be weaker than the paste; however, further research is needed to verify these hypotheses.

Conclusions

This chapter described the early age flexural strength behavior of paste, mortar, and concrete. A linear relationship was shown to exist between the flexural strength and nonevaporable water (i.e., degree of hydration) for paste specimens. However, as aggregates are added to the paste (i.e., mortar and concrete), a bilinear relationship is observed. This bilinear behavior occurs because the paste dominates failure at very early ages; however, at later ages the aggregates dominate the failure response. The rate of development of elastic modulus differs from that of flexural strength. Aggregates do not affect the elastic modulus in the same way they affect the flexural strength development.

Acknowledgments

The authors gratefully acknowledge support received from the National Science Foundation under grant no. 0134272, a career award granted to J. Weiss. Any opinions, findings, conclusions, or recommendations expressed in this material are those of the author(s) and do not necessarily reflect the views of the National Science Foundation (NSF). This work was conducted in the Charles Pankow Concrete Materials Laboratory, and the authors gratefully acknowledge the support that has made this laboratory and its operation possible.

References

1. A. G. A. Saul, "Principles Underlying the Steam Curing of Concrete at Atmospheric Pressure," *Mag. Concr. Res.,* **2** [6] 127 (1951).
2. K. M. Alexander and J. H. Taplin, "Concrete Strength, Cement Hydration and the Maturity Rule," *Austral. J. Appl. Sci.,* **13**, 277 (1962).
3. N. J. Carino, "The Maturity Method"; pp. 101–144 in *CRC Handbook on Nondestructive Testing of Concrete.* Edited by V. M. Malhotra and N. J. Carino. CRC Press, 1991.
4. H. P. Freiesleben and E. J. Pederson, "Maturity Computer for Controlled Curing and Hardening of Concrete," *Nordisk Betong,* **1**, 19 (1977).
5. C. Graveen, J. Weiss, J. Olek, and T. Nantung, "Implications of Maturity and Ultrasonic Wave Speed Measurements in QC/QA for Concrete"; pp 23–32 in *Seventh International Symposium on Brittle Matrix Composites, Poland.*
6. G. V. Guinea, K. Al-Sayed, C. G. Rocco, M. Elices, and J. Planas, "The Effect of Bond between the Matrix and the Aggregates on the Cracking Mechanism and Fracture Parameters of Concrete," *Cem. Concr. Res.,* **32** [12] 1961–1970 (2002).
7. M. J. Aquino, Z. Li, and S. P. Shah, "Mechanical Properties of the Aggregates and Cement Interface," *Adv. Cem. Based Mater.,* **2** [6] 211–223 (1996).
8. K. Wu, B. Chen, W. Yao, and D. Zhang, "Effect of Coarse Aggregate Type on the Mechanical Properties of High-Performance Concrete," *Cem. Concr. Res.,* **10** [31] 1421–1425 (2001).
9. A. Krishnan, "Durability of Concrete Containing Fly Ash or Slag Exposed to Low Temperature at Early Ages," thesis dissertation, Purdue University, 2002. Pp. 67–68.
10. P. Olken and F. S. Rostasy, "A Practical Planning Tool for the Simulation of Thermal Stresses and for the Prediction of Early Thermal Cracks in Massive Concrete Structures"; pp. 289–296 in *Thermal Cracking in Concrete at Early Ages.* Edited by R. Springenschmid. E&FN Spon, London, 1994.
11. P. N. Balaguru and S. P. Shah, *Fiber-Reinforced Cement Composites.* McGraw-Hill, New York, 1992. P. xii.

Drying Stresses and Internal Relative Humidity in Concrete

Zachary C. Grasley, David A. Lange, and Matthew D. D'Ambrosia

While concrete material technology has advanced significantly over the last 50 years in terms of mechanical and fluid properties, shrinkage cracking persists as a major problem without a complete solution. This chapter provides details on the mechanisms for microscale shrinkage and the development of shrinkage stress on the bulk scale in drying concrete, and explains how relative humidity measurements may be useful for quantifying shrinkage stresses. Recent research focused on measuring the internal relative humidity of concrete and current systems for measuring restrained stress development are summarized. Autogenous shrinkage models incorporating internal relative humidity measurements are outlined as are similar models that predict stress and strain gradients in drying concrete using relative humidity. Research has improved our understanding of how drying shrinkage stress gradients initiate microcracking on the drying surfaces of concrete structures and pavements.

Introduction

Many of the durability issues that plague concrete pavements and structures are dependent on the free ingress or movement of water. Cracking in concrete can enhance the ability of moisture to move freely in concrete, increasing the risk for durability problems such as corrosion, sulfate attack, and alkali-silica reaction.

Concrete cracking is often induced by internal stresses rather than an externally applied load. Internally induced stresses (i.e., material-induced stresses) are generated when a change in the equilibrium volume of a material is restrained. A number of material-induced volume changes contribute to stress accumulation and, ultimately, cracking in concrete. These include external drying shrinkage, internal drying (autogenous) shrinkage, thermal dilation, carbonation shrinkage, and crystallization pressure.

Internal and external drying shrinkage and thermal dilation are fundamentally related to the internal relative humidity (RH). This paper will discuss in detail the fundamental mechanisms behind RH and drying shrinkage stresses (internal and external drying) in concrete. In particular, the basis for the Kelvin and Laplace equations in early age shrinkage stress development is presented. Researchers have presented results utilizing the Kelvin-Laplace relationship for early age shrinkage research.[1-3] The complete derivation of both the Kelvin and Laplace equations is included in this review based on the usefulness of this relationship in shrinkage modeling.

Research projects involving internal RH measurements in concrete have recently increased. This is due in part to improvements in measurement techniques to quantify the internal RH. A number of different methods are currently in use to measure the internal RH. These methods are discussed along with any inherent advantages or disadvantages.

For cracking to occur as a result of drying shrinkage, some degree of restraint must also be present. The presence of restraint initiates a complex interaction between shrinkage and tensile creep. Much of the early work regarding shrinkage and creep of hardened concrete can be found in Ref. 4, while more recent modeling and experimental methods are discussed in this review.

The state of the art in measuring bulk stress due to external restraint is reviewed. The effect of internal and external restraint and creep will be discussed as will a simple model that relates the internal RH to the stress that develops in drying concrete. The model has been used to predict stress gradients associated with drying gradients, while similar modeling has been developed to predict autogenous shrinkage.[1,5]

While this paper is focused primarily on the physicochemical aspects of drying shrinkage mechanisms, stresses, and their relation to internal RH, a broader approach to early age volume change, restraint, and stress was published by Bissonnette et al. in *Materials Science of Concrete VI.*[6]

External Drying Shrinkage

Concrete is a porous material that undergoes a volume reduction upon the removal of pore water. Removal of the pore water by external drying occurs when the equilibrium RH of the ambient environment is lower than the internal RH. Moisture vapor diffuses from within the material to the surface, forcing more pore water to vaporize to maintain equilibrium. Based on thermodynamic equilibrium, at the drying front pores larger than about 50 nm will tend to empty first, followed by progressively smaller pores. Once pores have been emptied to a diameter of about 50 nm, curved capillary menisci and microscale shrinkage stresses will develop.[7] At this point, the internal RH will begin to decrease as well. At the drying front, a certain amount of drying can occur (and thus some weight loss) before microscale shrinkage stresses develop, since the larger pores will not have curved capillary menisci.

External drying shrinkage in concrete has traditionally been attributed to three mechanisms: capillary surface tension, disjoining pressure, and sur-

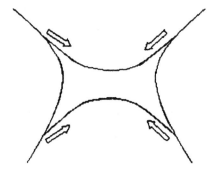

Figure 1. Misconception of surface tension shrinkage mechanism acting at the liquid/vapor interface.

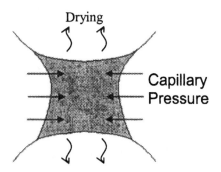

Figure 2. Negative pressure (underpressure) in pore solution associated with formation of menisci. (After Ref. 7.)

face free energy.[7] Whether a particular mechanism is active is dependent on the internal RH within the material. At mid to high internal RH, capillary surface tension and disjoining pressure are active. Below about 45% RH, capillary menisci are not stable, so capillary surface tension is not active below this level.[7] At lower RH, drying shrinkage is typically attributed to disjoining pressure and surface free energy. This paper will focus on the capillary surface tension mechanism and the disjoining pressure mechanism since they are active in the range of internal RH that a typical concrete structure will experience. Surface free energy will be discussed only briefly.

Capillary Surface Tension

Once the larger pores have emptied and pores with a diameter of about 50 nm begin to empty, curved capillary menisci will develop at the pore fluid/vapor interface. As the pore diameter of the still-saturated pores decreases, the surface curvature of the meniscus increases. It is tempting to mistakenly attribute the shrinkage stress associated with capillary tension to forces on the surface of the pore fluid pulling on the pore walls, as shown in Fig. 1. Although induced by surface tension, the shrinkage stress is a reduction in the pore fluid pressure relative to the balancing attractive forces between the solid microstructure,[8] depicted in Fig. 2.

The reduction in pore fluid pressure occurs as a requirement for static (mechanical) equilibrium caused by the change in surface curvature of the pore fluid. Figure 3(a) shows the vapor pressure, fluid pressure, and surface tension on a capillary meniscus in static equilibrium. Figure 3(b) presents a

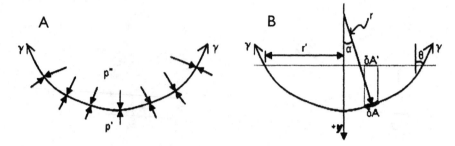

Figure 3. (A) Static equilibrium in capillary menisci. (B) Diagram for resolving forces in y direction.[9]

useful diagram for resolving forces in the vertical direction in a small, partially saturated pore. In the element δA, the force exerted normal to the element surface is the difference between the vapor (p') and fluid (p'') pressures multiplied by the area, that is, $(p''-p')\delta A$. Along the circular plane indicated by the horizontal dashed line, the force on the element $\delta A'$ is equal to $(p''-p')\cos \alpha \delta A$, which is the same as $(p''-p')\delta A'$. The sum of such vertical forces across the entire horizontal plane (dashed line) is simply the difference in the vapor pressure and the fluid pressure multiplied by the area of a circle (perpendicular to the plane within which the diagram lies) with radius r'. Thus, the sum of the forces in the y direction due to differences in fluid and vapor pressure is given by $(p''-p')\pi(r')^2$. The surface tension, γ, occurs along the entire perimeter, L, of the horizontal plane (dashed line). The additional vertical force component supplied by the surface tension is then

$$-\gamma \cos\theta \int_0^{2\pi r'} \delta L \tag{1}$$

Integration and substitution of r'/r for $\cos\theta$ yields

$$-\gamma \left(\frac{r'}{r}\right) 2\pi r' \tag{2}$$

Vertical mechanical equilibrium requires that the sum of forces in the y direction is equal to zero. This implies that

$$(p'' - p')\pi(r')^2 - \frac{-2\gamma}{r}\pi(r')^2 = 0 \qquad (3)$$

Reduction of Eq. 3 leads to the Laplace equation, which describes the mechanical equilibrium requirements for a spherical meniscus as

$$p'' - p' = \frac{2\gamma}{r} \qquad (4)$$

where p'' is the vapor pressure, p' is the pore fluid pressure, γ is the surface tension of water, and r is the average radius of curvature of the meniscus.[9] A more thorough version of the preceding derivation can be found elsewhere.[9]

Since the vapor pressure remains relatively constant compared to the large changes in the pore fluid pressure, the Laplace equation can be approximated as simply the development of an underpressure, σ, in the pore fluid of concrete as

$$\sigma = \frac{2\gamma}{r} \qquad (5)$$

As a result of this "tension" in the pore fluid, a balancing compressive stress develops in the solid microstructure, inducing microscale shrinkage.[6]

Figure 4 shows the negative pressure developed in the pore fluid at RH between 50 and 100%. Since capillary menisci are unstable below about 45% RH,[7] the capillary surface tension mechanism is active only above this level.

Disjoining Pressure

Not all liquid moisture in the pore structure of concrete is free to vaporize, adsorb, or desorb. A certain portion of the pore space in concrete consists of pores with diameters less than 30 Å,[10] and water molecules cannot freely adsorb in these small pores.[8] However, attractive van der Waals forces exist between the solid nanostructure and water molecules, which allow the presence of adsorbed water in regions where it would otherwise be hindered. The "swelling" pressure of this adsorbed water is often referred to as the disjoining pressure, and can be attributed to the double-layer repulsion and structural force of the adsorbed water.[11]

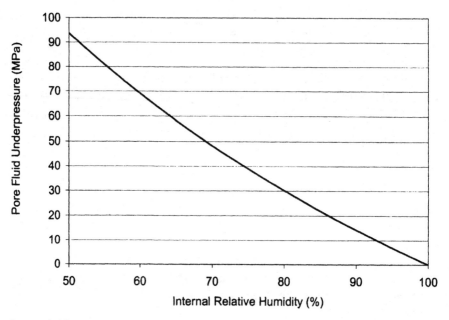

Figure 4. Negative pressure (underpressure) developed in pore fluid with changing internal RH.

Shrinkage stresses induced by the mechanism commonly referred to as disjoining pressure are caused by a reduction in the pressure pushing the solid microstructure apart relative to the attractive forces binding it together. Stresses develop in the solid microstructure to balance the reduction in disjoining pressure, resulting in microscale shrinkage. Therefore, the shrinkage mechanism for disjoining pressure is analogous to the mechanism for shrinkage due to capillary tension. In order to separate the mechanisms, it is necessary to qualify disjoining pressure as the force produced by adsorbed water only in areas where adsorption is hindered.[8]

Solid Surface Free Energy

Liquids such as water have surface tension at the liquid/vapor interface, and solids also have surface tension at their interface with other materials. If a solid is immersed in a like material, the surface tension is zero.[8] If it is immersed in water vapor, the solid surface tension of C-S-H is based on the concentration of water molecules in the air (i.e., partial vapor pressure).[8] In this manner, the solid surface tension is dependent on the thickness of the adsorbed layers of water. As the partial vapor pressure is reduced and hence the adsorbed thickness is decreased, the solid surface tension of C-S-H increases. This creates a net compression of the solid, resulting in micro-scale shrinkage.

When all of the free water is removed from the pore structure of hard-ened concrete, thin layers of adsorbed water remain. Further drying involves the removal of this layer of one or two molecules of thickness, which results in significant changes in the solid surface tension.[8] It is gener-ally assumed that changes in the solid surface tension of C-S-H do not become significant until lower RH (<50%), where only one or two adsorbed layers are present,[7] although some evidence does exist that solid surface tension contributes to shrinkage at higher RH as well.[12]

Internal Drying (Autogenous) Shrinkage

Autogenous shrinkage can be thought of as a special class of drying shrink-age where the moisture is removed from the pore structure through internal chemical and physical reactions rather than external drying. The hydration reaction that occurs when portland cement comes in contact with water leads to an overall volume reduction such that the volume of the products is less than the volume of the reactants. Once percolation of the solid phase occurs, the volume reduction leads to the formation of porosity as water is further consumed in the hydration reaction. Moisture will tend to be con-sumed from larger pores first, followed by successively smaller pores.

When high-performance concrete utilizing low water/cement (w/c) ratio and mineral admixtures is used, the larger pores are completely emptied as all of the water is required for hydration. Once pores smaller than 50 nm begin to empty, the same mechanisms that cause external drying shrinkage induce autogenous shrinkage. A typical concrete is already about 80% hydrated at 28 days, so autogenous shrinkage is primarily an early age problem. This type of shrinkage has drawn increased interest in recent years due to the use of lower w/c ratio concretes and mineral admixtures that consume additional water in a reaction to form secondary C-S-H. The formation of secondary C-S-H and the fine particle size of some mineral admixtures decrease the average pore size. This may increase meniscus curvature and thus increase autogenous shrinkage.

Relative Humidity

Relative humidity is the ratio of the partial vapor pressure (the measured vapor pressure) to the vapor pressure over a flat surface of water in a sealed container at a given temperature. This vapor pressure over a flat surface of water in a sealed container is not the maximum vapor pressure possible, although it is often termed the saturation pressure. Air can accommodate more water vapor than is typically present. In other words, it is not the capacity of the air that limits the maximum vapor pressure. In air surrounding small, convex water surfaces, such as minute droplets, the equilibrium vapor pressure is greater than what is defined as the saturation vapor pressure (i.e., RH > 100% is possible).[9,13] Equilibrium vapor pressure is the vapor pressure when the condensation and evaporation rates are equal.[14] Because the rate of condensation is always proportional to the partial vapor pressure, the evaporation rate controls the equilibrium vapor pressure.[14] Changes in temperature, water surface curvature, or dissolved salt ion concentration in the water all change the evaporation rate and thus change the equilibrium vapor pressure.

The reduction in the vapor pressure (and thus RH) due to dissolved salts in the pore water is important in concrete given the presence of potassium and sodium in cements. The presence of salt ions in the pore fluid lowers the equilibrium RH because the dissolved salt ions are surrounded by water molecules that are held by a strong ionic bond.[13] These bound water molecules are thus prevented from evaporation. By reducing the water molecules available for evaporation, the evaporation rate and the equilibrium RH are reduced. The effect that different dissolved salts and different concentrations have on the equilibrium RH can be investigated using Raoult's law. A derivation of Raoult's law can be found elsewhere.[1,15]

The surface tension of water also plays a role in controlling the evaporation rate. In the surface layer, each water molecule is attracted to the surrounding water molecules. On a convex surface (e.g., a small droplet), the water molecules have contact with fewer neighboring molecules than on a flat surface of water. The reduction in attractive forces binding each molecule to the liquid water reduces the energy required for vaporization. This increases the evaporation rate and subsequently increases the equilibrium vapor pressure.

Relationship between Relative Humidity and Drying Shrinkage Stresses

The previous section noted that when convex water surfaces are present, the equilibrium partial vapor pressure is greater than for a flat surface of water, which means an increase in RH beyond 100%. The opposite occurs when a concave surface is present as in the small micropores of concrete. The concave surface reduces the evaporation rate, which leads to a lower equilibrium RH. This was first explained thoroughly by Lord Kelvin, and allows the internal RH in concrete to be related to the meniscus curvature and, through the Laplace equation, to the pore fluid underpressure.

As mentioned previously, the surface curvature of capillary pore water creates an underpressure in the pore fluid as expressed in the Laplace equation (Eq. 4) due to the requirements for mechanical equilibrium. As physicochemical equilibrium must also be satisfied,

$$\mu' \quad \mu'' \quad \mu^\gamma \tag{6}$$

where μ' is the pore water chemical potential, μ'' is the vapor chemical potential, and μ^γ is the surface layer chemical potential.[9] A shift from one equilibrium condition to another can be shown as

$$\mu' \quad \mu'' \tag{7}$$

for physicochemical equilibrium and

$$\delta p'' \quad \delta p' = \delta\left(\frac{2\gamma}{r}\right) \tag{8}$$

for mechanical equilibrium (from the Laplace equation). Also, the Gibbs-Duhem equations state that

$$s' \ T \quad v' \ p' \quad ' \quad 0 \tag{9}$$

and

$$s'' \ T - v'' \ p'' + \mu'' \quad 0 \tag{10}$$

where s' and s'' are the molar entropies of the liquid and vapor phases, respectively, and v' and v'' are the molar volumes of the liquid and vapor phases, respectively. At a constant temperature, Eqs. 7, 9, and 10 may be combined to yield

$$v'\delta p' = v''\delta p'' \tag{11}$$

Combining this with Eq. 8 results in

$$\delta\left(\frac{2\gamma}{r}\right) = \frac{v' - v''}{v'}\delta p''$$ (12)

If the molar volume of the liquid, v', is neglected relative to the vapor molar volume, v'', and the vapor is treated as a perfect gas, then Eq. 12 can be written as

$$\delta\left(\frac{2\gamma}{r}\right) = \frac{-RT}{v'}\left(\frac{\delta p''}{p''}\right)$$ (13)

where R is the universal gas constant and T is the temperature in K. Integrating both sides of Eq. 13 from $1/r = 0$ to $1/r$ and from $p'' = p_{sat}$ to p'' leads to the Kelvin equation:

$$\frac{2\gamma}{r} = -\ln\left(\frac{p''}{p_{sat}}\right)\frac{RT}{v'}$$ (14)

where p_{sat} is the saturation vapor pressure at the given temperature. Since the ratio between the partial vapor pressure, p'', and the saturation vapor pressure, p_{sat}, can be expressed as RH (from 0 to 1) Eq. 14 can be written as

$$\frac{2\gamma}{r} = \frac{-\ln(RH)RT}{v'}$$ (15)

where γ is the surface tension of water, r is the average radius of meniscus curvature, RH is the internal RH, R is the universal gas constant, T is the temperature in Kelvin, and v' is the molar volume of water. A more thorough derivation of the Kelvin equation can be found elsewhere.[9] By combining the simplified Laplace (Eq. 5) and Kelvin (Eq. 15) equations, the underpressure developed in the pore fluid can be related to the internal RH as

$$\sigma = \frac{-\ln(RH)RT}{v'}$$ (16)

where all variables are as previously defined. An important factor to recognize in the derivation of the Kelvin and Laplace equations is that the pore radius, surface roughness, and contact angle of the meniscus and the pore wall do not affect the validity of the relationship between the RH and the underpressure developed. The pore radii are a controlling variable for both the internal RH and the underpressure through the mean radii of meniscus curvature. The internal RH and the underpressure are both dependent on changes to the mean radius of meniscus curvature, and any change in underpressure will be reflected by a corresponding change in internal RH.

It is not possible to quantify the absolute disjoining pressure (i.e., adsorbed water disjoining pressure). However, the change in pressure can be quantified using the Kelvin-Laplace equation since the change in stress of the adsorbed water must be the same as the change in stress for the evaporable water.[8] Therefore, at RH above ~45%, the Kelvin-Laplace equation accounts for both the shrinkage stress associated with the surface tension and the disjoining pressure mechanisms. As T. C. Powers stated, "To evaluate the capillary tension is to evaluate also the change in disjoining pressure."[8] As a result, it may be convenient to discuss surface tension and the disjoining pressure as a single mechanism (e.g., cumulative disjoining pressure) in discussions involving shrinkage stresses where the internal RH is above 45%, particularly when internal RH measurements are involved. In this case, drying shrinkage is primarily a result of reductions in the cumulative disjoining pressure relative to the attractive forces binding the microstructure together.

Figure 5 diagrams the relationship between internal RH and cumulative disjoining pressure and illustrates how a change in a parameter that affects underpressure correspondingly affects the internal RH. Issues such as microstructure roughness, which may be a source of error in determining pore radii distribution from internal RH measurements, are not a source of potential error in determining the cumulative disjoining pressure from internal RH measurements.

When considering internal RH measurements, one must consider the reduction caused by the internal salt concentration. Even when considering a simple a reduction in the internal RH from drying, it is possible that the salt concentration changes during drying. Figure 5 denotes the pore solution salt concentration as the only factor that affects the internal RH without affecting the capillary pressure in parallel.

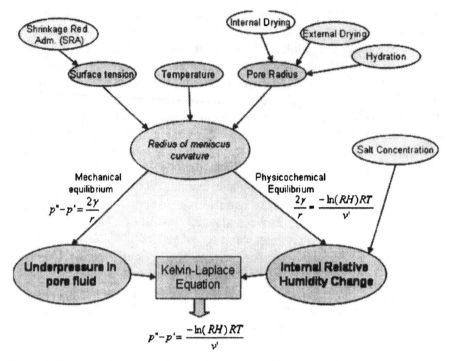

Figure 5. Diagram of factors affecting the internal RH and cumulative disjoining pressure in concrete.

Measuring Internal Relative Humidity

The ability to use the Kelvin-Laplace relationship for modeling of shrinkage stresses and strains is dependent on the ability to measure the internal RH. Recent improvements in sensors and other devices designed to measure internal RH along with the increasing importance of autogenous shrinkage (which cannot be characterized using traditional methods such as weight loss) has encouraged many researchers to investigate the RH within concrete or hardened cement paste.

Persson has recently studied the RH reduction (self-desiccation) in sealed high-performance concrete in relation to the autogenous shrinkage.[16] RH measurements were performed on crushed samples of concrete that were placed in glass tubes with a diameter of 25 mm. The samples were stored for at least 1 day, and then a dewpoint meter was placed tightly into the glass tubes. The RH was monitored for 1 day for each measurement. The autogenous shrinkage and RH were monitored for a period of at least 3

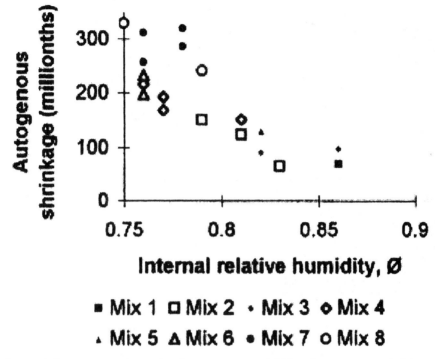

Figure 6. Long-term relationship between internal RH and autogenous shrinkage. (Reprinted with permission from Ref. 16.)

years, and a good correlation was observed between the long-term RH and autogenous shrinkage, as shown in Figure 6.[16]

Jensen and Hansen[2] have also investigated the internal RH reduction associated with autogenous shrinkage, but in hardened cement paste. The autogenous RH change was measured using a Rotronic DT hygroscope equipped with WA-14TH and WA-40TH measuring cells and thermostatically controlled. After initial set (8.5 h), the hardened cement paste was crushed and transferred to the measuring cells. More recently, Lura[1] measured self-deiscation using the same configuration.

Loukili et al.[17] researched the autogenous shrinkage and RH change in "ultra-high-strength" concrete (i.e., 150–400 MPa compressive strength). The internal RH was measured using a Vaisala Rotronic hygrometric probe. A 90 mm diameter × 180 mm height specimen was cast with a wooden stick placed in the middle of the circular cross section. At an age of 1 day, the specimen was demolded and sealed, and the stick was removed and

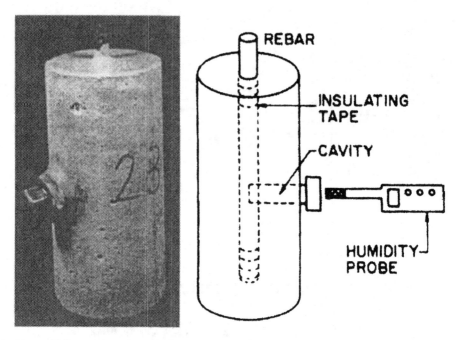

Figure 7. Experimental setup for measuring the internal RH near a reinforcing bar. (Reprinted with permission from Ref. 18.)

replaced with the probe. The probe was left in the measurement port throughout the experiment.

Jensen and Hansen measured autogenous internal RH reductions to about 90% after 7 days in moderately high-performance cement. Cement pastes with 0–10% cement replacement by silica fume and 0.35 w/c were tested. Lura tested two cement pastes, one with 76% replacement of portland cement with blast furnace slag, and both with 5.2% replacement of cement with silica fume and w/c = 0.37. Similar to Jensen and Hansen, Lura measured autogenous RH reductions to about 90% after 7 days. Loukili et al. measured autogenous RH reductions to below 75% RH after 10 days in "ultra-high performance" concrete. The w/c ratio for their test was 0.20, and 24% silica fume by weight of cement was added.

Andrade et al.[18] measured the internal RH in a concrete specimen to study the relationship between the corrosion of a reinforcing bar and the internal RH. Figure 7 illustrates the setup used to measure the internal RH. The cylindrical concrete specimens were 7.5 cm in diameter and 15 cm in height. A reinforcing bar was cast vertically in the cylinders. The specimens

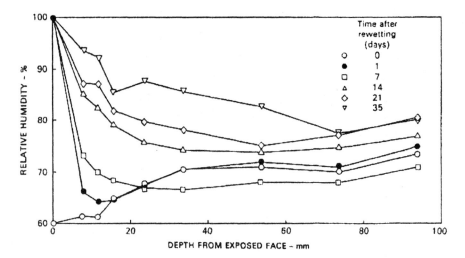

Figure 8. Internal RH profile in dried concrete exposed to rewetting.[19]

were cured in water for 3 days, and then stored in a climate-controlled chamber. Holes were drilled in the side of the specimens for inserting RH probes. Two different instruments were used to measure the internal RH: a hygrometric sensor and a portable Vaisala hygrometer. It was noted that the Vaisala hygrometer had no appreciable drift after 1 year and had a quick response time. Andrade et al. did find that at high RH it took over 24 h for the hygrometer to stabilize. The long stabilization time was avoided by storing the hygrometer in an environment with a similar RH to that within the concrete.

Other researchers have investigated the internal RH gradient in concrete. Parrott[19] measured the RH profiles within drying concrete and dried concrete that had been rewetted. Concrete cubes (100 mm) were cast in a mold containing eight removable steel plugs placed at increasing distances from the top (drying) surface. The plugs were removed at 70 h after casting, creating measurement ports for RH. The RH was measured using a small, capacitive RH probe. Figure 8 shows the internal RH gradient in dried concrete exposed to rewetting.

McCarter et al.[20] also measured the RH gradient in concrete exposed to drying. Solid PVC rods were cast into a concrete cylinder at depths of 15, 25, 35, and 45 mm from the drying surface. After demolding, the rods were removed to create cavities for RH measurement. RH measurements were made by placing a Rotronic M1 humidity probe into each measurement

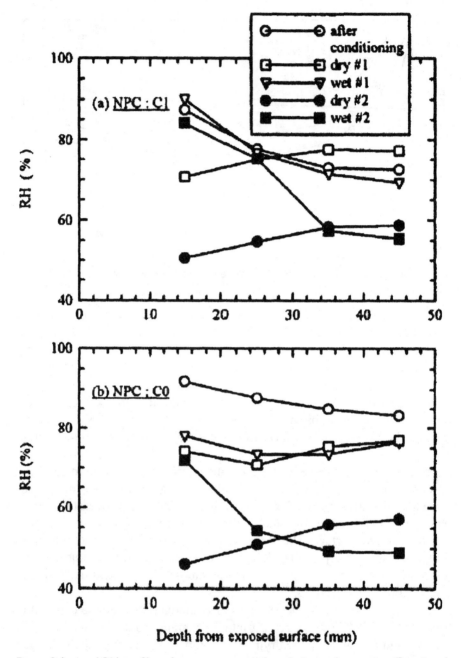

Figure 9. Internal RH profiles of concrete exposed to drying and rewetting. (Reprinted with permission from Ref. 20.)

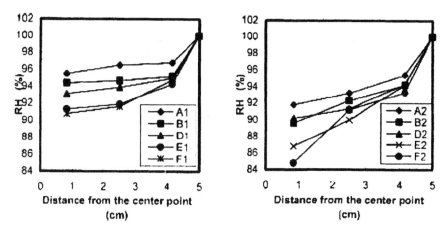

Figure 10. Internal RH gradient in (A) concrete and (B) silica fume–modified concrete that has been cured under water for 2 years (5 cm from center is surface). (Reprinted with permission from Ref. 21.)

cavity for a period of 10 min. Because of the short time allowed for equilibration of the RH within the measurement cavity after inserting the probe, the authors noted that the profiles shown in Fig. 9 may indicate only the relative change in RH rather than the absolute values.

Another method to measure the internal RH gradient was implemented by Yang.[21] Concrete cylinders with a diameter of 100 mm and a length of 250 mm were cast. After 2 years of storage in water, 24 10-mm slices were sawn from the cylinder using a wet saw. The pieces were considered to be taken from three separate layers (eight slices from each). Slices from the same layer were crushed together and placed in a small glass tube, which was sealed with a rubber plug. Protimeter dewpoint meters were used to measure the RH within the tubes. The RH sensors were inserted into each glass tube and left to equilibrate for at least 22 h before reading the measurement. Figure 10 shows the internal RH gradient in concrete cured under water for 2 years, where 5 cm from the center of the specimen represents the surface of the cylinder.

Lange and Shin[22] and Altoubat and Lange[23] have also measured the internal RH gradients in concrete. Altoubat and Lange measured the internal RH gradient using an RH probe and measurement ports in sealed concrete in an investigation of drying creep. Figure 11 shows the uniform profile and the overall reduction in internal RH in the sealed concrete due to self-desiccation. The uniform RH profile indicated that no stress gradient

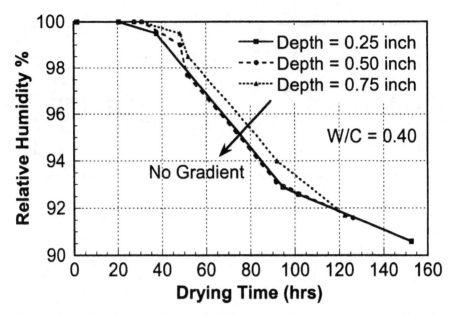

Figure 11. Uniform relative humidity profile in sealed concrete experiencing self-desiccation.[23]

Figure 12. Digital RH sensor (~20 mm in length).

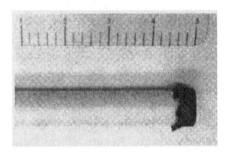

Figure 13. RH sensor packaged for embedding in fresh concrete.

was present, and, as a result, drying creep due to surface microcracking was discarded.

Grasley et al.[24,25] have also measured the internal RH gradient in concrete due to drying. A measurement system[26] using small, embeddable, dig-

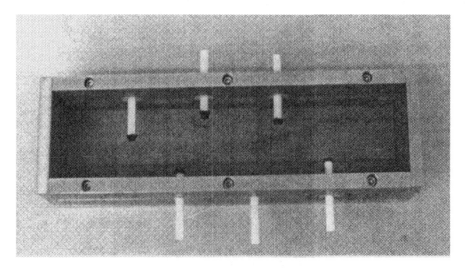

Figure 14. Mold for casting prism for measuring RH gradient.

ital RH sensors manufactured by Sensirion was used. The sensor, approximately 20 mm in length, is shown in Fig. 12. The sensor is packaged in a small plastic tube with one end sealed and the opposite end covered with GoreTex, as shown in Fig. 13. The GoreTex allows water vapor to pass through, but blocks any liquid water from entering the tube. This allows the sensor to be cast directly into fresh concrete. To measure the drying gradient, six sensors were cast directly into the fresh concrete using a special mold (Fig. 14). Internal RH measurements began immediately, and at an age of 1 day the specimen was demolded and the top and bottom sealed using adhesive-backed aluminum foil. This created symmetric drying conditions from the sides. The sensors were cast at depths of 6.4, 12.7, 19.0 (two sensors at this depth), 25.4, and 38.1 mm from the drying surface of the prism, which had a 76.2 × 76.2 mm cross section.

Shrinkage Stresses

When discussing drying shrinkage stresses, one must distinguish between microscale and bulk stresses. Microscale stresses induced by a reduction in the cumulative disjoining pressure are present whenever concrete is dried. Tensile shrinkage stresses on the bulk scale are present only when the concrete is restrained. Restraint of shrinkage on the bulk scale can be associated with aggregates, a drying gradient, or an "external" restraint, such as reinforcing steel or boundary conditions.

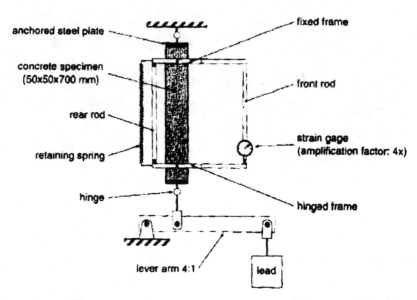

Figure 15. Dead load system for measuring constant load tensile creep. (Reprinted with permission from Ref. 27.)

The state of the art in quantifying stress associated with externally restrained shrinkage involves a uniaxial frame that is capable of applying a simulated, fully restrained load. These systems allow determination of the stress accumulation and creep relaxation under restrained condition. Passive systems, such as ring tests and passive test frames, are not included in this review because they are unable to simulate full restraint and quantify the resulting stresses.

Measurement of Average Stress Accumulation under Full External Restraint

The measurement of restrained shrinkage stress in concrete is critical to determining susceptibility to early age cracking. Conventional methods for measuring tensile creep under sustained loads, such as the experiment by Bissonnette and Pigeon shown in Fig. 15, were not capable of providing full restraint.[27] However, in the past 15 years, new testing techniques to measure restrained shrinkage stress have been developed.

Paillère et al.[28] used a uniaxial system to measure the stress developed due to restrained shrinkage. The system was developed at Laboratorie Cen-

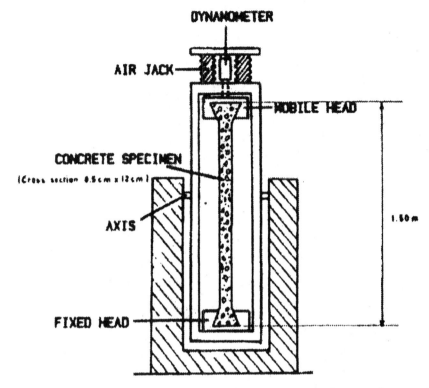

Figure 16. Cracking frame developed at Laboratorie Central des Ponts et Chaussées.[28]

tral des Ponts et Chaussées. A uniaxial specimen with flared ends was cast into a frame that applied a restraining force by means of an air pump, as shown in Fig. 16. Tensile stress was then measured with a load cell. Deformation was monitored and the load was applied manually to produce a restrained condition. This test was performed both vertically and horizontally, depending on the age of the specimen. It was determined that a vertical test was problematic owing to the dead load and fragility of the specimen.

Bloom and Bentur[29] developed a similar system in which an electronic step motor was used to apply the restraining load. Two flared-end specimens were measured for simultaneous determination of free shrinkage and stress development. Creep was calculated as the difference in strain accu-

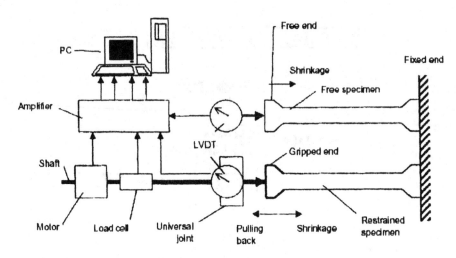

Figure 17. Restrained test used by Kovler.[30]

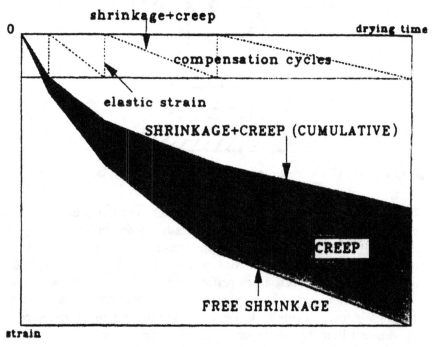

Figure 18. Schematic representation for simulation of restraint and determination of creep.[30]

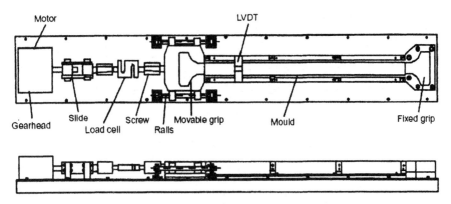

Figure 19. Restrained test developed by Pigeon et al.[36]

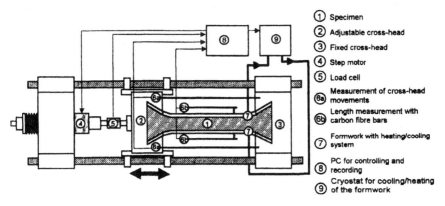

① Specimen
② Adjustable cross-head
③ Fixed cross-head
④ Step motor
⑤ Load cell
⑥ₐ Measurement of cross-head movements
⑥ᵦ Length measurement with carbon fibre bars
⑦ Formwork with heating/cooling system
⑧ PC for controlling and recording
⑨ Cryostat for cooling/heating of the formwork

Figure 20. TSTM developed by Springenschmid et al.[37]

mulation between the two specimens. Kovler[30] further modified this system to include a closed-loop computer control system, and measured deformation with LVDT sensors instead of conventional dial gauges. A schematic of this testing configuration is shown in Fig. 17.

When the deformation reached a predefined threshold, a restraining force was applied automatically to deform the specimen to its original length. A schematic of this testing philosophy is shown in Fig. 18. Many researchers have since adopted this method of simulating full restraint in a laboratory test. This experimental device has been used to investigate drying creep, autogenous shrinkage, and internal curing with lightweight aggregates.[31-35]

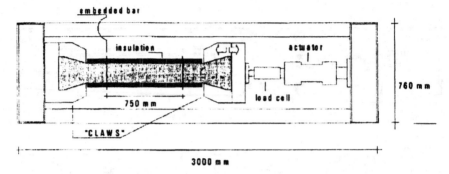

Figure 21. Testing device developed by van Breugel et al. (Reprinted with permission from Ref. 38.)

A test developed by Pigeon et al.,[36] based on Kovler's system, measures the stress due to restrained autogenous shrinkage. This experiment also uses a computer-controlled loading system. Deformation is measured using a direct current displacement transducer. A diagram of this experiment is shown in Fig. 19.

Springenschmid et al.[37] developed the Temperature Stress Testing Machine (TSTM) shown in Fig. 20 to measure the tensile stress in concrete due to the heat of hydration and external temperature change. Attached to one end of a uniaxial concrete specimen is an adjustable crosshead. A computer controlled step motor applies a load to control the deformation of the concrete specimen as it reaches a threshold of 0.001 mm (0.00004 in.).

Van Breugel and de Vreis developed a TSTM similar to that of Springenschmid et al., except that it uses a hydraulic actuator to apply load. This system is shown in Fig. 21. The device was used in conjunction with an autogenous deformation testing machine to optimize high-performance concrete mixture proportions based on creep and shrinkage performance. This device has been used to study the effect of curing temperature and type of cement on early age shrinkage of high-performance concrete.[38] Bjøntegaard and Sellevold[39] have used a similar TSTM to study the effect of temperature on the autogenous deformation of HPC.

Altoubat and Lange[40] developed the system currently in use at the University of Illinois at Urbana-Champaign, shown in Fig. 22. The 1000 × 75 × 75 mm (39 × 3 × 3 in.) specimen size allows testing of concrete with 25 mm (1 in.) coarse aggregate. The applied load is generated using a servo-hydraulic actuator with high load stability that is capable of load application up to 90 kN (20 kip). An extensometer anchored in the specimen is used to

Figure 22. Uniaxial restrained stress device developed by Altoubat and Lange.[40]

avoid grip–specimen interaction, which caused inaccurate strain measurements in preliminary tests.[41] Deformation is measured with an LVDT and fed into a closed-loop system that controls the applied load to the specimen, which is measured with a load cell. A threshold value of 0.005 mm (8 μm/m) is used to simulate restraint. This value was determined experimentally to be the minimum effective value within the limitations of the measuring equipment and environmental conditions. This system has been used to study drying creep, fiber-reinforced concrete, and the effects of shrinkage-reducing admixtures on creep and stress development.[42–44]

Modeling Free Shrinkage Stress Gradients Caused by External Drying

Bulk-scale stresses in free shrinkage specimens are a result of shrinkage restraint through aggregates and a moisture gradient. Closer to the drying surface, the concrete will exhibit a faster reduction in internal RH. Figure 23 illustrates the development of a moisture (RH) gradient in a concrete prism exposed to drying conditions.

The restraint caused by the moisture gradient is a result of the inner core of the material preventing the outer surface from shrinking to its equilibrium, zero-stress position. Likewise, the high equilibrium shrinkage of the surface material forces the inner core to deform beyond its equilibrium position. The equilibrium, zero-stress shrinkage can be referred to as the "unrestrained shrinkage."[45] To approach measurements of true unrestrained drying shrinkage, one must reduce the specimen cross section significantly

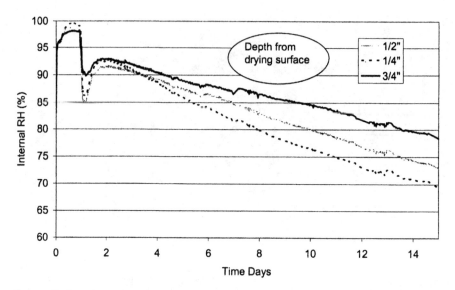

Figure 23. Development of moisture gradient in concrete prism exposed to 50% RH and 23°C. The sharp drop at 1 day is due to evaporative cooling upon form removal.

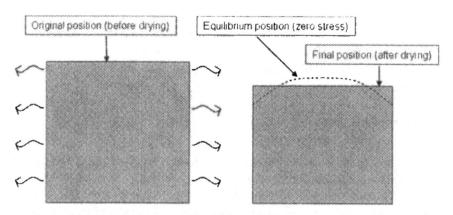

Figure 24. Mechanism for bulk scale drying shrinkage stresses in free shrinkage specimen where translational symmetry is imposed.

(e.g., 1–3 mm thickness), which has been done with hardened cement paste.[12,46] While small concrete specimens may exhibit measurable differential shrinkage across the cross section,[47] the surface curvature or curling is certainly not enough to fully alleviate shrinkage stress development as a result of internal restraint.

In certain cases the end conditions of a free shrinkage specimen require that each element of the cross section exhibit bulk shrinkage at the same rate. Figure 24 shows the effect of this "translational symmetry"[48] on symmetrically drying concrete. Figure 24 also implies that the center of the specimen is in compression while the outer surfaces are in tension.

The difference between the equilibrium (zero-stress) position and the position required by translational symmetry is composed of two strain components: creep strain and some remaining strain that has an associated stress. Assuming specimen geometry allows for a plane stress approximation, the bulk measured "free" shrinkage strain, ε_T, at any location across the specimen cross section to be expressed as

$$\varepsilon_T = \varepsilon_{sh} + \varepsilon_{cr} + \varepsilon_{el} \tag{17}$$

where ε_{sh} is the equilibrium free shrinkage, ε_{cr} is the strain relaxed by creep, and ε_{el} is the remaining strain that has an associated stress. If we approximate this strain as completely linear elastic, the stress, σ_{el}, at any point across a free shrinkage specimen cross section can be calculated as

$$\sigma_{el} = (\varepsilon_T - \varepsilon_{sh}) E_{concrete} - \varepsilon_{cr} E_{concrete} \tag{18}$$

where $E_{concrete}$ is the elastic modulus of the concrete. The potential stress in the absence of creep, σ_{pot}, which will be useful later in this discussion, can be expressed as

$$\sigma_{pot} = (\varepsilon_T - \varepsilon_{sh}) E_{concrete} \tag{19}$$

where all the variables are as previously defined.

The linear strain of an elastic solid with spherical pores exerting a pressure can be determined using[49]

$$\varepsilon = \sigma \left(\frac{1}{3k} - \frac{1}{3k_0} \right) \tag{20}$$

where σ is the pressure (or underpressure) exerted in the pores, k is the bulk modulus of the porous solid, and k_0 is the bulk modulus of the solid skeleton. The underpressure, σ, can be determined from the Kelvin-Laplace equation using RH measurements. The result of Eq. 20 is the equilibrium free shrinkage strain, ε_{sh}, for a saturated sample. For the partially saturated case, Bentz et al.[3] modified Eq. 20 to include a saturation factor as

$$\varepsilon_{sh} = \varepsilon = \sigma S \left(\frac{1}{3k} - \frac{1}{3k_0} \right) \tag{21}$$

where S is the saturation factor. Research on shrinkage of partially saturated porous Vycor glass demonstrated that the modified equation gives results comparable to experimental measurements.[3]

Equation 21 can be used to determine the equilibrium shrinkage of hardened cement paste (HCP). A couple of models are available for converting HCP strain to concrete strain based on aggregate volume fraction and stiffness.[50,51] However, converting the HCP strain to a concrete strain in this manner neglects any bulk scale stresses in the paste caused by aggregate restraint. Therefore, the equilibrium shrinkage of the concrete can be approximated by simply multiplying Eq. 21 by the paste volume fraction in each mixture as

$$\varepsilon_{sh} = \sigma S \left(\frac{1}{3k} - \frac{1}{3k_0} \right) \left(\frac{v_p}{v_t} \right) \tag{22}$$

where v_p is the volume of paste and v_t is the total volume of concrete. In effect, Eq. 22 dilutes the shrinkage of the paste based on the volume of aggregates but does not ignore the elastic strain (and thus the stress) in the paste imposed by the presence of aggregates.

To determine the creep strain, ε_{cr}, an existing model may be used. Previous work has implemented the B3 model[52] to predict the microstructural basic and drying creep.[24] Although the B3 model was developed for bulk measured creep, it was applied to this model by treating the rectangular specimen cross section as a series of finite-width elements. Each of these elements had a potential stress, σ_{pot}, associated with the local equilibrium shrinkage, ε_{sh}. The compliance components of the B3 model, $C_0(t,t')$ and $C_d(t,t',t_0)$, were determined for each individual element, where $C_0(t,t')$ is the basic creep compliance and $C_d(t,t',t_0)$ is the drying creep compliance. Instead of using an estimate of the average internal RH as was intended in the drying creep compliance function in the B3 model, the actual measured internal RH at each discrete location was used. The total stress relaxation from creep at any point across the cross section was determined as

$$\sigma_{cr} = \left[C_0 \left(t,t' \right) + C_d \left(t,t',t_0 \right) \right] \sigma_{pot} E_{concrete} \tag{23}$$

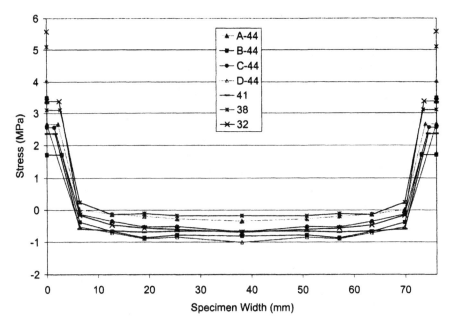

Figure 25. Stress distribution in free shrinkage concrete specimen exposed to symmetric drying for 4 days.

where σ_{pot} is the potential stress, which is equal to $(\varepsilon_T - \varepsilon_{sh})E_{concrete}$ as shown in Eq. 19.

With the equilibrium shrinkage strain, ε_{sh}, and the stress relaxed by creep calculated, the remaining stress at any point across the specimen cross section can be determined as

$$\sigma_{el} = (\varepsilon_T - \varepsilon_{sh}) E_{concrete} - \left[C_0 (t,t') + C_d (t,t',t_0) \right] (\varepsilon_T - \varepsilon_{sh}) E^2_{concrete} \quad (24)$$

Figure 25 shows the model predicted bulk stress distribution across a 72 × 72 mm concrete cross section exposed to 50% ambient RH and 23°C symmetric drying conditions from the sides. The key in Fig. 25 lists a number of concrete mixtures modeled, where the number in the mixture designation represents the w/cm ratio. Two specimens with identical cross sections and volume/surface ratio were cast: a prism specimen, which was used to measure the RH gradient using a measurement system developed at the University of Illinois at Urbana-Champaign,[26] and a uniaxial specimen, as shown in Fig. 22. The specimens were demolded at 1 day and exposed to drying for an additional 4 days. The model predicted that the material was

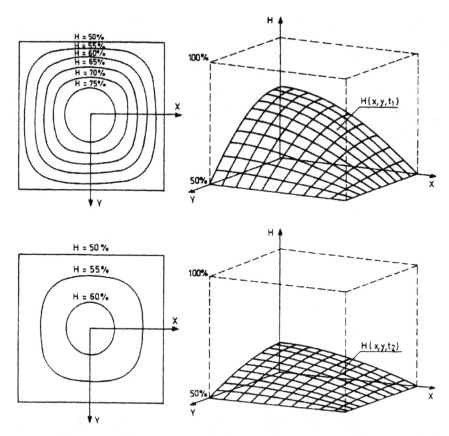

Figure 26. Moisture gradients in drying prism at two different stages of drying modeled using diffusion theory. (Reprinted with permission from Ref. 45.)

under compression in the moist core and in tension near the surface. The flat portions of the curves at the very outer surface represent areas where the predicted tensile stresses exceeded the measured tensile strength of the material. In the model, these areas have been treated as damaged zones (microcracking) that behave plastically. While no effort was made to validate the presence of surface microcracks in this study, other researchers have verified their presence on the surface of drying concrete or HCP.[46,53,54] A more complete experimental analysis of the free shrinkage stress distribution model described herein can be found elsewhere.[24,25]

Wittmann and Roelfstra[45] modeled the shrinkage stress gradient in a free shrinkage externally drying concrete prism. The moisture gradient was

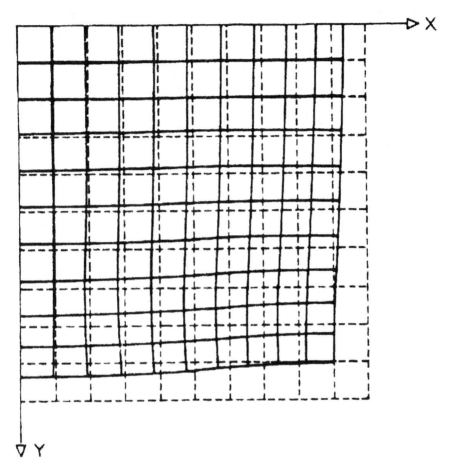

Figure 27. Model predicted shrinkage strains. (Reprinted with permission from Ref. 45.)

modeled using linear diffusion theory. The equilibrium unrestrained free shrinkage was determined using an equation developed by Klug.[55] Klug measured the shrinkage of thin hardened cement paste plates between 0.5 and 5 mm thickness to extrapolate the zero thickness (unrestrained) shrinkage at different RH. Klug derived a relationship between the RH and unrestrained shrinkage as

$$\varepsilon_u = ah + b \qquad (25)$$

where h is the RH, ε_u is the unrestrained shrinkage, and a and b are material

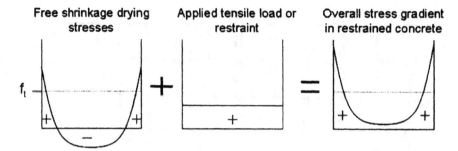

Figure 28. Superposition of free shrinkage drying stresses and stresses due to external restraint.

parameters based on concrete composition and maturity. Wittman used the moisture gradient developed through the linear diffusion equation to approximate the internal RH profile as shown in Fig. 26.

Using Eq. 25 to determine the unrestrained shrinkage and rate theory to describe the creep relaxation, the stress and strain gradients were modeled using finite element analysis. The predicted two-dimensional strain profile is shown in Fig. 27.

Modeling Fully Restrained Shrinkage Stress Gradients Caused by External Drying

Analysis of the stress gradient in concrete specimens that are fully restrained externally (i.e., zero measured bulk shrinkage) is similar to the analysis of free shrinkage stress gradients. The only difference between the free shrinkage stress gradients and the restrained shrinkage stress gradients is an applied stress that is constant across the specimen cross section. The applied stress is that which is required to maintain zero shrinkage. This idea of superposition is illustrated in Fig. 28. In order to model the stresses in fully restrained specimens exposed to external drying conditions, the stress required to maintain zero shrinkage must be measured.

The uniaxial test frames described earlier are designed to determine the stress accumulation under full restraint. Superimposing this applied restraint stress, $\sigma_{applied}$, to the free shrinkage stress gradient yields

$$\sigma_{el} = (\varepsilon_T - \varepsilon_{sh}) E_{concrete} - [C_0(t,t') + C_d(t,t',t_0)](\varepsilon_T - \varepsilon_{sh}) E_{concrete}^2$$
$$+ \sigma_{applied} \tag{26}$$

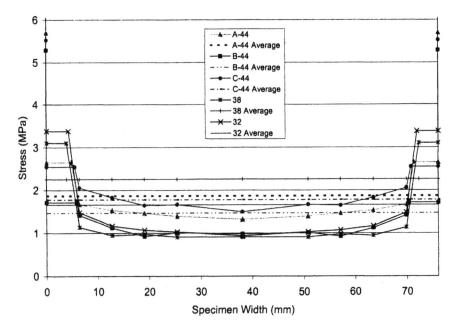

Figure 29. Stress distribution in fully restrained concrete specimen exposed to symmetric drying for 4 days.

which allows the stress at any point across the cross section of a fully restrained specimen to be calculated.

Figure 29 shows the modeled stress distribution in fully restrained uniaxial specimens exposed to symmetric drying for 4 days. The drying conditions were 50% RH and 23°C. The same mixtures are plotted in Fig. 29 as for the free shrinkage stress distributions in Fig. 25. The horizontal lines in Fig. 29 are the externally applied stresses for each material. As in Fig. 25, the flat portions of the curves near the surface represent areas where the tensile stress exceeds the tensile strength of the material. As one would expect, these predicted damaged zones penetrate farther into the fully restrained specimens than the free shrinkage specimens.

Evidence for Stress Gradient

Both the modeled free shrinkage stress gradients and the fully restrained shrinkage stress gradients predict zones near the drying surface where the tensile stresses in the material exceed the tensile strength. Since large, visible macrocracks are typically not present on the surface of free shrinkage

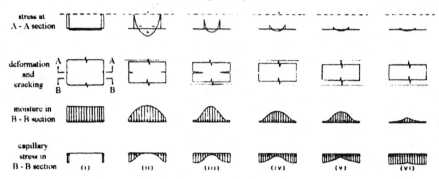

tensile strength of the paste

stress at
A - A section

deformation
and
cracking

moisture in
B - B section

capillary
stress in
B - B section

(i)　(ii)　(iii)　(iv)　(v)　(vi)

Figure 30. Evolution of surface microcracks, internal stresses, moisture, and capillary stress in a thin hardened cement paste specimen. (Reprinted with permission from Ref. 46.)

concrete, this implies that microcracking is present. To verify the presence of microcracking on the surface, research on plain cement paste is appropriate since any microcracking would be due to self-restraint from the moisture gradient rather than aggregate restraint.

Hwang and Young[46] used light microscopy to investigate the occurrence of microcracks on the surface of thin free shrinkage cement paste specimens. Pastes with 0.4 w/c were cast in an 11.5 × 11.5 × 1.15 mm mold. After set, the specimens were demolded and stored in lime water for a period of 10 weeks. The specimens were sliced to 1 mm thickness and stored briefly in a 100% RH environment until testing began. Detection of microcracks using light microscopy initiated as soon as the specimen was removed from the 100% RH environment. It was found that the surface microcracks were present and progressed as shown in Fig. 30, where Stage I is 10 min after exposure to drying, Stage II is 10 min to 4 h, Stage III is 4 h to 1 day, Stage IV is 1–2 days, Stage V is 2–4 days, and Stage VI is after 4 days of drying. Figure 30 also shows qualitatively that the moisture gradient (and hence the stress gradient) is most severe shortly after exposure to drying. Microcracks were found to close up as time progressed.

Bisschop[54] cast cement paste specimens with w/c = 0.45. The 40 × 40 × 160 mm prisms were demolded at 24 h, then stored in water for 6 days. The prisms were allowed to dry for 16 h, at which time the surface cracks were examined using optical microscopy with ultraviolet light. A digitized image of the microcracks observed using this technique is shown in Fig. 31,

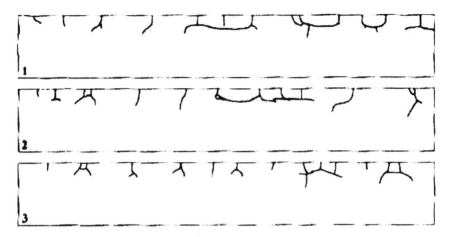

Figure 31. Surface microcracking in hardened cement paste due to drying gradient.[54]

where the cross sections are 20 × 135 mm. The deepest microcracks observed at 16 h were 8.1 mm deep. The model for concrete free shrinkage stress gradients discussed in this paper predicts that the zone of microcracking extends to 2–6 mm from the drying surface for a variety of mixtures after 4 days of drying. Hwang and Young[46] reported that microcracks close up over time since the moisture gradient becomes less severe as drying continues. In addition, the severity of the stress gradient is inherently dependent on the size of the specimen as well as the volume/surface ratio. Considering these issues, the microcrack depths observed by Bisschop are in fair agreement with the model prediction.

Other researchers have developed models that predict microcracking due to moisture gradients. Bazant and Raftshol[48] modeled the stresses associated with a drying gradient as well as the resulting microcracking in hardened cement paste. Assuming a linear elastic material and linear diffusion theory, they predicted that a difference of only ~2% internal RH between the surface of a specimen and the center of the specimen would result in microcracking. Since it is known that cementitious materials are viscoelastic rather than linear elastic, it is reasonable to assume that some of the stresses associated with the drying gradient are relaxed because of creep. Therefore, the internal RH difference required to initiate microcracking in a real material is probably higher than 2%.

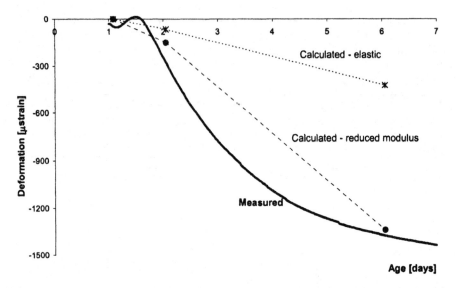

Figure 32. Predicted versus measured autogenous shrinkage of hardened cement paste.[5]

Modeling Autogenous Shrinkage Strains in Sealed Concrete

In free shrinkage specimens exposed to internal drying alone, severe moisture gradients do not develop. This is because internal drying occurs at a relatively constant rate across the specimen cross section. In most laboratory-sized specimens the reaction temperature distribution is not severe enough to cause a noticeably higher rate of hydration-induced moisture consumption at different locations across the cross section. Therefore, stress gradients do not develop in sealed concrete free shrinkage specimens. As a result,

$$\varepsilon_T = \varepsilon_{sh} \qquad (27)$$

at any point across the cross section. This allows the bulk measured free shrinkage strain to easily be predicted by a single internal RH measurement using the same approach as shown in the stress gradient model. Other researchers have used this relationship to predict the autogenous shrinkage of hardened cement paste[5] and concrete.[1]

Lura et al.[5] modeled the autogenous shrinkage of hardened cement paste using different cement types. The bulk scale paste shrinkage, ε_T, was pre-

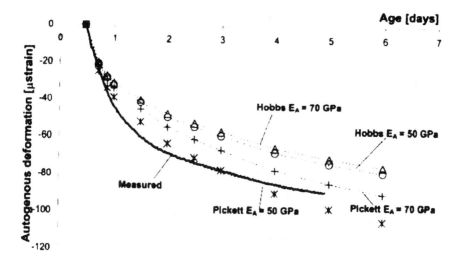

Figure 33. Agreement between model prediction and measured autogenous shrinkage varying the elastic modulus of the aggregates.[1]

dicted using Eq. 21 with autogenous shrinkage, $\varepsilon_T = \varepsilon_{sh}$. The negative pore fluid pressure, σ, was determined with the Kelvin-Laplace equation through internal RH measurements. The predicted deformation assuming linear elasticity was significantly lower than the measured deformation. Lura et al. hypothesized that short-term microscale creep could account for the additional deformation observed in measured autogenous shrinkage specimens. The effect of this short-term creep was accounted for by using a reduced elastic modulus, as shown in Fig. 32.

Lura[1] has also modeled the shrinkage of concrete. The paste shrinkage was modeled in the same fashion as with the previous reference. To predict the shrinkage of the concrete materials, the paste shrinkage was converted to concrete shrinkage using Pickett's[50] and Hobbs's[51] models. Reasonable agreement was found with measured autogenous shrinkage, with the relative accuracy depending on which model was used as well as which value was used for aggregate modulus of elasticity. The results are shown in Fig. 33.

Summary

The mechanisms of internal and external drying shrinkage have been discussed in detail. Drying shrinkage mechanisms are fundamentally related to the internal RH within the concrete pore structure. The basis of the physicochemical equilibrium of RH was outlined as well as current methods and applications of RH measurement.

Capillary stresses are the major driving force behind early age drying shrinkage. Drying shrinkage mechanisms are the same whether driven by external drying or internal drying. Internal drying of sealed samples will typically reduce RH to a minimum of about 75% before free water for continued hydration is completely consumed.

Shrinkage of concrete will cause stresses on the bulk scale when restraint is present. Internal restraint will be present if moisture or temperature gradients exist. External restraint is often present in real structures, and can be simulated using a uniaxial test frame. A model was described that generates stress distributions across free shrinkage specimens or externally restrained specimens. In sealed specimens (internal drying only), the same model principles may be used to predict the stress-free bulk shrinkage.

Moisture gradients are important because the stress gradients they induce can cause surface microcracking. This microcracking will reduce durability and may eventually encourage the formation of macrocracks. Models that calculate stress gradients could eventually be expanded to include macrocrack formation by incorporating fracture mechanics concepts. Reliable models of drying shrinkage behavior are important for developing new materials and improved concrete mixture designs.

References

1. P. Lura, "Autogenous Deformation and Internal Curing of Concrete," Ph.D. thesis, Delft University, 2003.
2. O. M. Jensen and P. F. Hansen, "Autogenous Deformation and Change of Relative Humidity in Silica Fume-Modified Cement Paste," *ACI Mater. J.*, **93**, 539–543 (1996).
3. D. P. Bentz, E. J. Garboczi, and D. A. Quenard, "Modelling Drying Shrinkage in Reconstructed Porous Materials: Application to Porous Vycor Glass," *Modelling Simul. Mater. Sci. Eng.*, **6**, 211–236 (1998).
4. Z. P. Bazant and F. H. Wittmann, eds. *Creep and Shrinkage in Concrete Structures*. Wiley, New York, 1982.
5. P. Lura, Y. Guang, and K. van Breugel, "Effect of Cement Type on Autogenous Deformation of Cement-Based Materials," ACI-SP no. TBD. American Concrete Institute, 2003.
6. B. Bissonette, J. Marchand, J. P. Charron, A. Delagrave, and L. Barcelo, "Early Age Behavior of Cement-Based Materials"; in *Materials Science of Concrete VI*. Edited by J. Skalny and S. Mindess. American Ceramic Society, Westerville, Ohio, 2001.
7. S. Mindess and J. F. Young, *Concrete*. Prentice-Hall, Englewood Cliffs, New Jersey, 1981.

8. T. C. Powers, "Session I: The Thermodynamics of Volume Change and Creep," *Materiaux Const.*, **1**, 487–507 (1968).

9. R. Defay, I. Prigogine, A. Bellemans, and D. H. Everett, *Surface Tension and Adsorption.* John Wiley & Sons, New York, 1966.

10. R. S. Mikahil and S. A. Selim, "Adsorption of Organic Vapors in Relation to Pore Structure of Hardened Portland Cement Pastes"; pp. 123–134 in *Symposium on the Structure of Portland Cement Paste in Concrete.* Highway Research Board Special Report 90. Washington, D.C., 1996.

11. C. F. Ferraris and F. H. Wittmann, "Shrinkage Mechanisms of Hardened Cement Paste," *Cem. Concr. Res.*, **17**, 453–464 (1987).

12. W. Hansen, "Drying Shrinkage Mechanisms in Portland Cement Paste," *J. Am. Ceram. Soc.*, **70**, 323–328 (1987).

13. A. W. Adamson, *A Textbook of Physical Chemistry*, 3rd ed. Academic Press, 1986.

14. G. Shortley and D. Williams, *Elements of Physics*, 5th ed. Prentice-Hall, 1971.

15. O. M. Jensen, "Appendix: Measurements and Notes"; in *Autogenous Deformation and RH Change — Self-Desiccation and Self-Desiccation Shrinkage* (in Danish). TR 285/93. Building Materials Laboratory, Technical University of Denmark, 1993.

16. B. Persson, "Experimental Studies on Shrinkage of High-Performance Concrete," *Cem. Concr. Res.*, **28**, 1023–1036 (1998).

17. A. Loukili, A. Khelidj, and P. Richard, "Hydration Kinetics, Change of Relative Humidity, and Autogenous Shrinkage of Ultra-High-Strength Concrete," *Cem. Concr. Res.*, **29**, 577–584 (1999).

18. C. Andrade, J. Sarria, and C. Alonso, "Relative Humidity in the Interior of Concrete Exposed to Natural and Artificial Weathering," *Cem. Concr. Res.*, **29**, 1249–1259 (1999).

19. L. J. Parrott, "Moisture Profiles in Drying Concrete," *Adv. Cem. Res.*, **1**, 164–170 (1988).

20. W. J. McCarter, D. W. Watson, and T. M. Chrisp, "Surface Zone Concrete: Drying, Absorption, and Moisture Distribution," *J. Mater. Civil Eng.*, **13**, 49–57 (2001).

21. Q. Yang, "Inner Relative Humidity and Degree of Saturation in High-Performance Concrete Stored in Water or Salt Solution for 2 Years," *Cem. Concr. Res.*, **29**, 45–53 (1999).

22. D. A. Lange and H.-C. Shin, "Early Age Stresses and Debonding in Bonded Concrete Overlays," *Transp. Res. Record*, **1778**, 174–181 (2001).

23. S. A. Altoubat and D. A. Lange, "The Pickett Effect at Early Age and Experiment Separating Its Mechanisms in Tension," *Mater. Struct.*, **34**, 211–218 (2002).

24. Z. C. Grasley, D. A. Lange, and M. D. D'Ambrosia, "Internal Relative Humidity and Drying Stress Gradients in Concrete"; pp. 349–363 in *Proceedings of the Engineering Conferences International.* Advances in Cement and Concrete IX. 2003.

25. Z. C. Grasley, "Internal Relative Humidity, Drying Stress Gradients, and Hygrothermal Dilation of Concrete," M.S. thesis, University of Illinois at Urbana-Champaign, 2003.

26. Z. C. Grasley, D. A. Lange, and M. D. D'Ambrosia, "Embedded Sensors for Measuring Internal Relative Humidity in Concrete," submitted to *Cem. Concr. Res.*

27. B. Bissonnette and M. Pigeon, "Tensile Creep at Early Ages of Ordinary, Silica Fume

and Fiber Reinforced Concretes," *Cem. Concr. Res.,* **25,** 1075–1085 (1995).

28. A. M. Paillère, M. Buil, and J. J. Serrano, "Effect of Fiber Addition on the Autogenous Shrinkage of Silica Fume Concrete," *ACI Mater. J.,* **86,** 139–144 (1989).

29. R. Bloom and A. Bentur, "Free and Restrained Shrinkage of Normal and High Strength Concretes," *ACI Mater. J.,* **92,** 211–217 (1995).

30. K. Kovler, "Testing System for Determining the Mechanical Behavior of Early Age Concrete under Restrained and Free Uniaxial Shrinkage," *Mater. Struct.,* **27,** 324–330 (1994).

31. K. Kovler, "New Look at the Problem of Drying Creep of Concrete under Tension," *J. Mater. Civil Eng.,* **11,** 84–87 (1999).

32. K. Kovler, "Interdependence of Creep and Shrinkage for Concrete under Tension," *J. Mater. Civil Eng.,* **7,** 96–101 (1995).

33. S. Igarashi, A. Bentur, and K. Kovler, "Autogenous Shrinkage and Induced Restraining Stresses in High-Strength Concretes," *Cem. Concr. Res.,* **30,** 1701–1707 (2000).

34. S. Igarashi, A. Bentur, and K. Kovler, "Stresses and Creep Relaxation Induced in Restrained Autogenous Shrinkage of High-Strength Pastes and Concretes," *Adv. Cem. Res.,* **11,** 169–177 (1999).

35. A. Bentur, S. Igarashi, and K. Kovler, "Prevention of Autogenous Shrinkage in High-Strength Concrete by Internal Curing Using Wet Lightweight Aggregates," *Cem. Concr. Res.,* **31,** 1587–1591 (2001).

36. M. Pigeon, G. Toma, A. Delagrave, B. Bissonnette, J. Marchand, and J. C. Prince, "Equipment for the Analysis of the Behaviour of Concrete under Restrained Shrinkage at Early Ages," *Mag. Concr. Res.,* **52,** 297–302 (2000).

37. R. Springenschmid, R. Breitenbücher, and M. Mangold, "Development of Thermal Cracking Frame and the Temperature-Stress Testing Machine"; pp. 137–144 in *Thermal Cracking in Concrete at Early Ages* (Proceedings of the 5th International RILEM Symposium, Munich). Edited by R. Springenschmid. E&FN Spon, 1995.

38. P. Lura, K. van Breugel, and I. Maruyama, "Effect of Curing Temperature and Type of Cement on Early-Age Shrinkage of High-Performance Concrete," *Cem. Concr. Res.,* **31,** 1867–1872 (2001).

39. Ø. Bjøntegaard and E. J. Sellevold, pp. 125–140 in *Effects of Silica Fume and Temperature on Autogenous Deformation of High Performance Concrete.* ACI SP-220. Edited by O. M. Jensen, D. P. Bentz, and P. Lura. 2004.

40. S. A. Altoubat, "Early Age Stresses and Creep-Shrinkage Interaction of Restrained Concrete," Ph.D. thesis, Department of Civil and Environmental Engineering, University of Illinois at Urbana-Champaign, 2000.

41. S. A. Altoubat and D. A. Lange, "Grip-Specimen Interaction in Uniaxial Restrained Test"; in *Concrete: Material Science to Application — A Tribute to S. P. Shah.* ACI SP-206. American Concrete Institute, 2002.

42. S. A. Altoubat and D. A. Lange, "The Pickett Effect at Early Age and Experiment Separating Its Mechanisms in Tension," *Mater. Struct.,* **34,** 211–218 (2002).

43. S. A. Altoubat and D. A. Lange, "A New Look at Tensile Creep of Fiber Reinforced Concrete"; in *ACI Special Publication on Fiber Reinforced Concrete.* Edited by N. Banthia. American Concrete Institute, 2003.

44. M. D. D'Ambrosia, S. A. Altoubat, C. Park, and D. A. Lange, "Early Age Tensile Creep

and Shrinkage of Concrete with Shrinkage Reducing Admixtures"; pp. 685–690 in *Creep, Shrinkage and Durability Mechanics of Concrete and other Quasi-Brittle Materials* (Proceeding of CONCREEP '01, Boston, August 13–15, 2001). Edited by F. Ulm, Z. Bazant, and F. H. Wittman. 2001.

45. F. H. Wittmann and P. E. Roelfstra, "Total Deformation of Loaded Drying Concrete," *Cem. Concr. Res.,* **10**, 601–610 (1980).

46. C.-L. Hwang and J. F. Young, "Drying Shrinkage of Portland Cement Pastes I. Microcracking During Drying," *Cem. Concr. Res.,* **14**, 585–594 (1984).

47. A. M. Alvaredo and F. H. Wittmann, "Shrinkage as Influenced by Strain Softening and Crack Formation, in Creep and Shrinkage of Concrete"; pp. 103–113 in *Thermal Cracking in Concrete at Early Ages* (Proceedings of the 5th International RILEM Symposium, Munich). Edited by R. Springenschmid. E&FN Spon, 1993.

48. Z. P. Bazant and W. J. Raftshol, "Effect of Cracking in Drying and Shrinkage Specimens," *Cem. Concr. Res.,* **12**, 209–226 (1982).

49. J. K. Mackenzie, "The Elastic Constants of a Solid Containing Spherical Holes," *Proc. Phys. Soc. B,* **683**, 2–11 (1950).

50. G. Pickett, "Effect of Aggregate on Shrinkage of Concrete and a Hypothesis Concerning Shrinkage," *ACI J.,* **52**, 581–590 (1956).

51. D. W. Hobbs, "Influence of Aggregate Restraint on the Shrinkage of Concrete," *ACI J.,* **71**, 445–450 (1974).

52. "Creep and Shrinkage Prediction Model for Analysis and Design of Concrete Structures — Model B3" (RILEM Draft Recommendation), *Mater. Struct.,* **28**, 357–365 (1995).

53. D. D. Higgins and J. E. Bailey, "A Microstructural Investigation of the Failure Behaviour of Cement Paste"; pp. 283–296 in *Proceedings of the Conference on Hydraulic Cement Pastes: Their Structure and Properties* (Sheffield, England). 1976.

54. J. Bisschop, "Drying Shrinkage Microcracking in Cement-Based Materials," Ph.D. thesis, Delft University, 2002.

55. P. Klug, "Kriechen, Relaxation und Schwinden von Zementstein," Ph.D. dissertation, TU Munchen, 1973.

9 781574 982107